Cryobiology

Cryobiology

Julian White

SYRAWOOD
PUBLISHING HOUSE
New York

Published by Syrawood Publishing House,
750 Third Avenue, 9th Floor,
New York, NY 10017, USA
www.syrawoodpublishinghouse.com

Cryobiology
Julian White

International Standard Book Number: 978-1-64740-098-9 (Hardback)

Cataloging-in-Publication Data

Cryobiology / Julian White.
p. cm.
Includes bibliographical references and index.
ISBN 978-1-64740-098-9
1. Cryobiology. 2. Cryopreservation of organs, tissues, etc. 3. Low temperatures.
4. Biology. I. White, Julian.
QH324.9.C7 C79 2022
571.464 5--dc23

TABLE OF CONTENTS

Preface VII

Chapter 1 What is Cryobiology? 1

Chapter 2 Processes in Cryobiology 12

- Cryosurgery 12
- Cryoablation 30
- Cryoimmunotherapy 40
- Cryoneurolysis 42
- Cryotherapy 44

Chapter 3 Cryopreservation 48

- Precision in Cryopreservation 59
- Cryoconservation of Animal Genetic Resources 77
- Cryopreservation of Testicular Tissue 88
- Semen Cryopreservation 101
- Oocyte Cryopreservation 103
- Ovarian Tissue Cryopreservation 107
- Embryo Cryopreservation 108
- Cryopreservation Methods 110
- Cryopreservation of Hematopoietic Stem Cells 118

Chapter 4 Cryogenics 129

- Cryogenic Fuel 133
- Cryogenic Energy Storage 136
- Cryocooler 138
- Cryogenic Milling 145
- Cryogenic Processor 145
- Cooling Bath 147

- Helium Cryogenics 150
- Cryogenic Seal 153
- Cryotank 156

Chapter 5 **Diverse Aspects of Cryobiology** **172**

- Cryonics 172
- Antifreeze Protein 172
- Hypothermia 177
- Cryobank 187
- Cryochemistry 196
- Cryochemical Synthesis 197
- Cryobiology and Organ Preservation 198

Permissions

Index

PREFACE

This book has been written, keeping in view that students want more practical information. Thus, my aim has been to make it as comprehensive as possible for the readers. I would like to extend my thanks to my family and co-workers for their knowledge, support and encouragement all along.

The biological branch which deals with the study of the effects of low temperatures on living beings in the cryosphere of the Earth is known as cryobiology. The biological material and systems are studied at moderately hypothermic conditions to cryogenic temperatures. Cryobiology primarily studies systems and materials such as organs, tissues, proteins, cells as well as whole organisms. The central areas of study in cryobiology are the study of cryosurgery, and cold-adaptation of microorganisms, animals and plants. It also studies cryopreservation of cells, gametes, tissues of animals and humans. It also deals with the preservation of organs under hypothetical conditions and physics of supercooling. This book provides significant information of this discipline to help develop a good understanding of cryobiology and related fields. While understanding the long-term perspectives of the topics, it makes an effort in highlighting their impact as a modern tool for the growth of the discipline. The book is appropriate for students seeking detailed information in this area as well as for experts.

.A brief description of the chapters is provided below for further understanding:

Chapter - What is Cryobiology?

Cryobiology is the branch of biology that deals with the effects of low temperature on the living organisms. It includes study of proteins, cells, tissues, organs, etc. when they are maintained at a temperature ranging from hypothermic condition to cryogenic condition. This is an introductory chapter which will briefly introduce all the significant aspects of cryobiology.

Chapter - Processes in Cryobiology

Cryobiology is a vast subject that includes numerous processes such as cryosurgery, cryoablation, cryoimmunotheraphy, cryoneurolysis, cryotherapy, cryosurgery, cryotransport, etc. The topics elaborated in this chapter will help in gaining a better perspective about various processes of cryobiology.

Chapter - Cryopreservation

Cryopreservation uses very low temperatures to preserve various living contents such as organelles, cells, tissues, extracellular matrix, organs, or any other biological constructs. This chapter delves into cryoconservation of animal genetic resources and cryopreservation of testicular tissue, semen, oocyte, etc. to provide an easy understanding of the subject.

Chapter - Cryogenics

Cryogenics refers to the study of materials and their behavioral aspects due to production and maintenance at extremely low temperature. Some of the elements of cryogenics are cryogenic fuel, cryogenic energy storage, cryocooler, cryogenic processor, cooling bath, etc. The topics elaborated in this chapter will help in gaining a better perspective about cryogenics.

Chapter - Diverse Aspects of Cryobiology

Cryobiology is a vast subject that has diverse aspects such as cryonics, cryoneurolysis, antifreeze protein, cryobank, cryochemistry, cryochemical synthesis, cryobiology and organ preservation, etc. The aim of this chapter is to explore these applications of cryobiology which are closely related to its progress.

Julian White

What is Cryobiology?

CHAPTER 1

Cryobiology is the branch of biology that deals with the effects of low temperature on the living organisms. It includes study of proteins, cells, tissues, organs, etc. when they are maintained at a temperature ranging from hypothermic condition to cryogenic condition. This is an introductory chapter which will briefly introduce all the significant aspects of cryobiology.

Cryobiology deals with life at low temperature. Living things must be able to adapt to the changing surface environment of the earth in order to preserve the existence of life itself. The principal effects of cold on living tissue are destruction of life and preservation of life at a reduced level of activity. Both of these effects are demonstrated in nature. Death by freezing is a relatively common occurrence in severe winter storms. Among cold-blooded animals, winter weather usually results in a coma-like sleep that may last for a considerable length of time. Therefore, the definition of cryobiology is to study living organisms at low temperature. In other words, cryobiology is the branch of biology involving the study of the effects of low temperatures on organisms (most often for the purpose of achieving cryopreservation).

In cryobiological applications, much lower temperatures are used are present in natural environments. Liquid nitrogen (at −196 °C or −320 °F) can either destroy living tissue in a matter of seconds or it can preserve it for years, and possibly for centuries, with essentially no detectable biochemical activity. The end result when heat is withdrawn from living tissue depends on processes occurring in the individual cells. Basic knowledge of the causes of cell death, especially during the process of freezing, and the discovery of methods which prevented these causes, have led to practical applications for long-term storage of both living cells and living tissues. In the industrial food area, the microorganisms used in cheese production can be frozen, stored, and transported without loss of lactic acid-producing activity. In the medical field, it is commonly known that whole blood or separated blood cells can be cryopreserved and stored for their valuable applications.

Water is the fundamental molecule of life. The biochemical constituents of a cell are either dissolved or suspended in water. Water is essential for the survival of all known forms of life; without an environment of water, life would not exist. Water has many distinct properties that are critical for the proliferation of life and these set it apart from other substances. It enables the proliferation of life by allowing organic compounds to react in ways that ultimately allow replication. Water is vital both as a solvent for many of the body's solutes and as an essential part of many metabolic processes within the body. Water is essential and central to these metabolic processes. Metabolic processes are affected by temperature. When the temperature falls, cells may slow down or stop all metabolic processes, and extremely low temperature may cause cell death.

During the physical process of freezing, water tends to crystallize in pure form, while the dissolved or suspended materials concentrate in the remaining liquid. In the living cell, this process is quite destructive. In a relatively slow-freezing process, ice first begins to form in the fluid surrounding the cells, and the concentration of dissolved materials in the remaining li quid increases. A concentration gradient is established across the cell wall, and water moves out of the cell in response to the osmotic force. As freezing continues, the cell becomes relatively dehydrated. Salts may concentrate to extremely high levels. In a similar manner, the acid–base ratio of the solution may be altered during the concentration process.

Dehydration can affect the gross organization of the cell and also molecular relationships, some of which depend on the presence of water at particular sites. Cellular collapse resulting from loss of water may bring into contact intracellular components that are normally separated to prevent any destructive interaction. Finally, as the ice crystals grow in size, the cell walls may be ruptured by the crystals themselves or by the high concentration gradients that are imposed upon the walls. To prevent dehydration, steps must be taken to stop the separation of water in the form of pure ice so that all of the cell fluids can solidify together.

Cryobiology is the core of fertility cryopreservation. The earliest application of fertility cryopreservation was in the storage of animal sperm cells for use in artificial insemination. The principal application for human fertility cryopreservation was also begun with sperm freezing, and then with embryo and oocyte as well as gonadal cryopreservation. Knowledge and medical achievement have steadily advanced in the field of fertility cryopreservation, especially with recent oocyte and ovarian tissue cryopreservation. These historic accomplishments in the application of the scientific method can provide overwhelming support for continuing on this path.

Basic Science of Cryobiology

Nature of Water

Water appears in nature in all three common states of matter: Vapor, liquid, and solid. Water is a tasteless, odorless liquid at standard temperature and pressure. The color of water and ice is, intrinsically, a very light blue, although water appears colorless in small quantities. Ice also appears colorless, and water vapor is essentially invisible as a gas. The maximum density of water occurs at 3.98 °C (39.16 °F). Water becomes even less dense upon freezing, expanding 9%. This causes an unusual phenomenon: ice floats upon water, and so organisms can live inside a partly frozen pond because the water on the bottom has a temperature of around 4 °C (39 °F). The boiling point of water is 100 °C (212 °F) at sea level and one atmosphere pressure. The freezing point of water is very close to 0 °C (32 °F) in the presence of nucleating substances, but in their absence it can be supercooled to −42 °C (−43.6 °F) before freezing. For most substances, freezing and melting points are approximately equal. Therefore, the melting point of ice at one atmosphere pressure is very close to 0 °C (32 °F). The melting point of water is relatively insensitive to change in pressure because the solid–liquid transition represents only a small change in volume.

The transition between liquid water and solid ice is one of the most commonly observed events in nature. As mentioned above, when water is cooled, it often is taken substantially below the freezing point before ice begins to form. This is because of the need for nucleation to occur before an ice crystal can begin to grow. Nucleation refers to the process by which a minimum crystal is formed, which can then expand. The continued expansion of the crystal is a process known as growth. When an ice nucleus begins to grow, any solutes that are present in the liquid will be excluded from this growing ice front. If the rate of crystal growth is faster than the rate at which diffusion of the particular solutes can carry them away from the ice front, then a concentration gradient will very quickly form in the liquid that surrounds the ice crystal. The concentrated solute will then lower the freezing point of the solution. When a certain amount of ice has formed, the solution at the interface will have a freezing point equal to the temperature of the interface. At this point, ice growth will be limited by diffusion of the solute away from the crystal. If the temperature is reduced to far below the melting point with supercooling speed, the solution may be prevented from reaching this situation of ice crystal nucleation and growth. If water is cooled sufficiently fast enough so that nucleation cannot occur, it is possible to avoid ice crystal formation; this process is known as vitrification.

Temperature Measurement

Thermometers measure temperature by using materials that change in some way when they are heated or cooled. In a mercury or alcohol thermometer, the liquid expands as it is heated and contracts when it is cooled, so the length of the liquid column is longer or shorter depending on the temperature. Modern thermometers are calibrated in standard temperature units such as Fahrenheit (F), Celsius (C), or Kelvin (K).

Celsius is converted to Fahrenheit by multiplying by 1.8 (or 9/5) and add 32, and to Kelvin by adding 273 (e.g. 37 °C is equivalent to 98.6 °F and 310 K).

Glass Transition Temperature

The glass transition temperature (T_g) is the temperature at which an amorphous solid becomes brittle on cooling or soft on heating. Glass transition is a pseudo-second phase transition in which a supercooling melt yields on cooling a glassy structure with properties similar to those of crystalline materials. Below T_g, amorphous solids in a glassy state, and most of their joining bonds are intact. It is important to note that T_g is a kinetic parameter and, therefore, parametrically depends on the melt cooling rate. Consequently, the slower the melt cooling rate, the lower the value of T_g. In addition, T_g depends on the measurement conditions, which are not universally defined.

At a certain temperature, the average kinetic energy of molecules no longer exceeds the binding energy between neighboring molecules, and growth of an organized solid crystal begins. Formation of an ordered system takes a certain amount of time since the molecules must move from their current location to energetically preferred points at crystal nodes. As the temperature falls, molecular motion slows down further and if the cooling rate is fast enough, molecules never reach their destination: the substance enters into dynamic arrest and a disordered glassy solid form. A full discussion of T_g requires an understanding of mechanical loss mechanisms of specific functional groups and molecular arrangements. The value of T_g is somewhat dependent on the time-scale of the imposed change in contrast to the melting point temperatures of crystalline materials. Time and temperature are interchangeable quantities when dealing with glasses, a fact often expressed in the time–temperature superposition principle. An alternative way to discuss the same issue is to say that a T_g is only a point on the temperature scale if the change is imposed at one particular frequency. Since T_g is dependent on the cooling rate as the glass is formed, the glass transition is not considered a true thermodynamic phase transition by many in the field.

The viscosity at T_g depends on the sample preparation (especially the cooling curve), the heating or cooling curve during measurement, and the chemical composition. Proteins possess a T_g value below which both anharmonic motions and long-range correlated motion within a single molecule are quenched. The origin of this transition is primarily a consequence of caging by glassy water, but it can also be modeled in the absence of explicit water molecules, suggesting that part of the transition reflects internal protein dynamics. Glass formation of water below the melting point can occur, usually through very rapid cooling or the introduction of agents that suppress the formation of ice crystals.

Vitrification

Vitrification is defined as the process of glass solidification of a liquid. The liquid is in a metastable state until it gets below a characteristic temperature, T_g, which is indicated by a sharp exothermic event. This heat loss occurs because of the loss of metastable clusters. Those clusters that have more energy than can be held by the bonds which they are able to form will oscillate for a short time and then disintegrate. Upon reaching T_g, the excess energy will be lost, thereby stabilizing the clusters. Once below T_g, the system is not merely a viscous liquid but is also a solid that is in a stable thermodynamic state. Achieving vitrification with pure water requires very small amounts and incredibly fast cooling. However, it is important to mention that vitrification can also occur in aqueous solution during slow freezing.

Aqueous Solutions

A solution is a mixture containing at least two kinds of pure substance. In most solutions, one material predominates and this is called the solvent, with the other compounds being called solutes. When the components of a solution are in different states of matter, the solvent is considered to be the one that does not undergo a change of state upon mixing. Aqueous solutions are important for cryobiology since the freezing of biological systems always involves solutions containing substances such as electrolytes, non-electrolytes, polymers, and so on. During the phase change that occurs with vitrification, the concentration and distribution of the solutions are altered, sometimes accompanied by irreversible chemical reactions.

Molarity and Molality

The composition of a solution is described by the concentration of its constitutions. There are two primary ways of expressing concentration: molarity and molality. Molarity is another term for concentration (M) and is the number of moles of solute in 1 liter of solution. Molality (m) is the number of moles of solute associated with 1000 g of solvent. Therefore, molarity is based on the volume of solution whereas molality is based on the weight of solvent. The difference becomes most noticeable when temperature effects are considered. Because the volume of liquids can expand or contract with changes in temperature, molarity can change with change of temperature. By comparison, the weight of solvent is constant with temperature, so molality gives a measure of concentration that is independent of temperature.

Solubility

A property of any particular combination of solute and solvent is the solubility of the solute in the solvent. This is the amount of solute that can be associated with a given amount of solvent in the context of a solution. Most solids show a well-defined saturation point in liquid when no more solid can be dissolved. The ratio of solute to solvent at this point defines the solubility. In general, solubility in liquids increases with temperature. Therefore, the solubility is a temperature-dependent property, although there are some exceptions. If a saturated solution is made at a certain temperature and then the temperature of the solution decreases, the solubility of the solution will be exceeded. The solution becomes supersaturated and exists in a metastable state. The solute will precipitate out of solution, usually forming crystals in the liquid. Even in crystalline form, solute molecules continually leave and join the crystal surface, going back and forth from solution to the solid phase. This has the effect of increasing the average crystal size, since small crystals have a high surface energy while large crystals have a small surface energy. The constant movement between crystal and solution tends to minimize the total surface energy of all the crystals present in the solution. The solubility is important for cryobiology because the solubility of solutes in extracellular and intracellular water will be changed following the change of temperature.

Therefore, changes in solubility during freezing may induce ice crystal formation and cause cell death.

Colligative Properties

The properties that solutions exhibit that arise from the behavior of the collection rather than from the behavior of individual components are called colligative properties. Vapor pressure, boiling point elevation, and freezing point depression are three such properties, depending upon the concentration of the solutions rather than the chemical properties of the constituents.

Osmosis

Water can be transported across semipermeable membranes separating compartments containing different concentrations of solutes. The membrane must be impermeable to the solute but permeable for water. This process is called osmosis and has enormous significance for living organisms. The most important and most widely occurring process for water transfer in and out of living cells is osmosis.

Cell Permeability

Changes in the extracellular osmotic pressure will create a situation in which a cell will attempt to attain equilibration by either gaining or losing water until there is no osmotic gradient across the plasma membrane. If the cell volume is measured as a function of time, then it can be seen that equilibrium is only achieved after a certain amount of time has elapsed. The kinetics of water movement out of the cell is determined by the physical structure of the membrane. With biologic al membranes, the phenomenological permeability is complicated. Since freezing and thawing introduce opportunities for osmotic swelling and shrinkage of cells, it may be important to know the tolerance of each type of cell for the osmotic pressure during a given cell type's response to exposure to low temperatures.

Cryoinjury

Some of the classic papers in the field of cryobiology describe the theories and the mechanisms of cryoinjury during cell freezing and thawing. These theories have made great contributions to developments and understanding in cryobiology. Cryoinjury has been successfully simulated by changing the concentration of solutes surrounding cells in suspension so as to simulate the changes in concentration that take place upon freezing and thawing when water is subtracted or added back to yield the original pre-freezing concentrations. Many theories and mechanisms have been proposed for cryoinjury, but none may exactly explain the nature of the phenomenon. For example, it is not surprising that survival rates can vary from cell type to cell type for the same cooling rate and freezing solution. Although there are mathematical models describing how to calculate appropriate cooling rates for avoiding intracellular ice formation, the theoretical predictions may not apply for all types of cell, particularly for aqueous solutions supplemented with cryoprotective additives. Some cryoprotectants reduce the injury of cells during freezing and thawing. Cryoprotectants are usually divided into two broad classes based on their ability to diffuse across cell membranes. Penetrating cryoprotectants are able to move across cell membranes whereas non-penetrating agents cannot.

Cryoprotectants

Discovery of Cryoprotectant Properties

Although a good survival rate of deep-frozen cells has occasionally been observed without a protective agent, a suitable cryoprotectant usually increases the survival rate. Usually the literature indicates that Polge et al. first reported that glycerol has cryo protective function to improve survival rate of frozen–thawed cells (chicken spermatozoa). However, the cryoprotective effect of glycerol was discovered much earlier than is usually stated. Although survived cells have been occasionally without a cryoprotectant, the presence of a suitable cryoprotectant usually increases the cell survival rates considerably. The discovery that glycerol, first, and later dimethyl sulfoxide protect eukaryotic cells against freezing damage marked the beginning of modern cryobiology.

Today, the most commonly used cryoprotectants in the field are glycerol, dimethyl sulfoxide, ethylene glycol, and propylene glycol. The cryoprotective action of each type of cryoprotective agent must be similar, but although many hypotheses have been proposed to explain their mechanism of action, it is still unclear what role they do actually play in the freezing or vitrification solutions. For example, Lovelock proposed that glycerol acted colligatively (altering the phase diagram of the solution) to reduce the high salt concentration that occurs during freezing. Later, phase diagrams were produced that described the mechanism of action of cryoprotectants, especially dimethyl sulfoxide. Phase diagrams are used to describe equilibrium situations in which two or more phases of matter exist together as pure substances or in solutions. In the freezing system, the primary component is water, and the entire system is a collection of compartments filled with an aqueous solution. As aqueous solutions are cooled, the water forms a crystalline solid that has almost no solubility for the solutes that were in the aqueous solution. As ice forms, the solutes will be confined to the remaining liquid phase, becoming more concentrated. Because this lowers the freezing point of the aqueous phase, the system can remain in equilibrium with a substantial unfrozen fraction. As cooling continues, the solubility limit of the solution will also be reached, leading to the precipitation of solutes. The ternary systems of glycerol– NaCl–water and dimethylsulfoxide–NaCl–water have been described. From these diagrams, it is clear that the solubility and eutectic behavior of a single solute can be altered significantly by the amount and type of additional solutes introduced into the system. It is also clear that the equilibrium between solids and liquid becomes increasingly complex as the number of components is increased. Therefore, the action of cryoprotectants can be described as lowering the freezing point and reducing/preventing ice crystal formation of aqueous (freezing) solutions.

The following compounds are commonly used cryoprotectants in field of cryobiology.

Glycerol

Glycerol, also known as glycerin or glycerine, is a sugar alcohol. It is a colorless, odorless, viscous, sweettasting liquid that is soluble in water and low in toxicity. Each glycerol molecule has a three-carbon chain, with a hydroxyl group (OH) attached to each carbon atom. The hydroxyl groups are responsible for making the substance highly soluble in water and hygroscopic. A hygroscopic substance is one that attracts water molecules from the surrounding environment. It has only slight solubility in organic solvents such as ethyl acetate and diethyl ether, and it does not dissolve in hydrocarbons. Its melting point is 18 °C (64.4 °F), and its boiling point is 290 °C (554 °F). Its surface tension is 64.00 mN/m at 20 °C, and it has a temperature coefficient of −0.0598 mN/(m K). The glycerol substructure is a central component of many lipids. Glycerol is useful for numerous applications. It is a common component of solvents for enzymatic reagents stored at temperatures below 0 °C as the presence of glycerol depresses the freezing temperature of the solution. Glycerol is compatible with other biochemical materials in living cells and is frequently used in cell preservation to reduce damage caused by ice crystal formation.

Dimethyl Sulfoxide

Dimethyl sulfoxide, also known as methyl sulfoxide or methylsulfinylmethane, is a clear and colorless li quid. The sulfur center in dimethyl sulfoxide is nucleophilic toward soft electrophiles and the oxygen is nucleophilic toward hard electrophiles. The methyl groups are somewhat acidic in character ($pK_a = 35$) because of the stabilization of the resultant carbanion by the S(O)R group, and so are deprotonated with strong bases such as lithium diisopropylamide and sodium hydride. Dimethyl sulfoxide is an import ant polar aprotic solvent that dissolves both polar and non-polar compounds and it is miscible in a wide range of organic solvents as well as water.

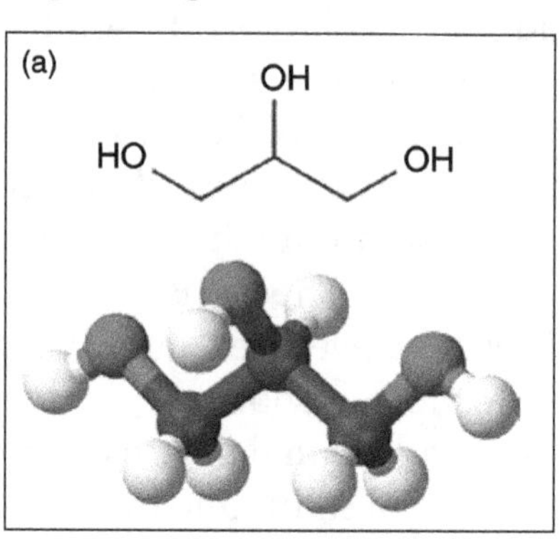

Glycerol.

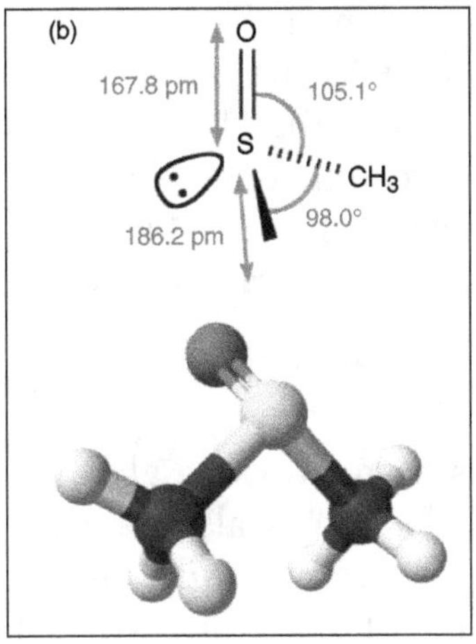

Dimethyl sulfoxide.

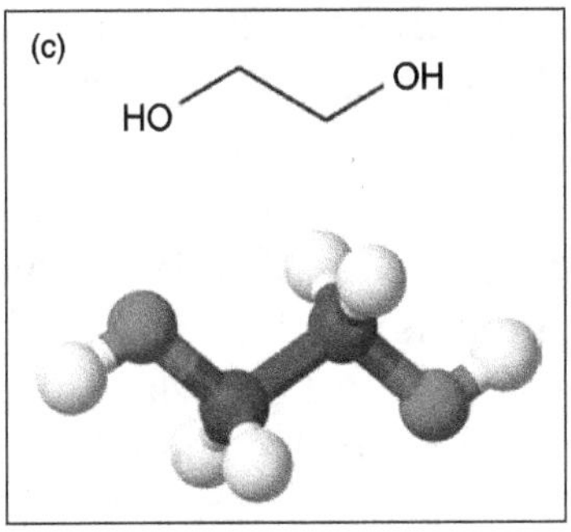

Ethylene glycol.

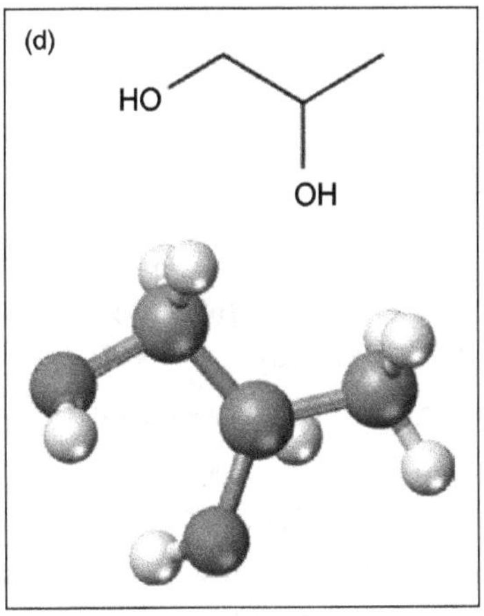

Propylene glycol.
The structure of cryoprotectants commonly used in cryobiology.

It dissolves a variety of organic substances, including carbohydrates, polymers, and peptides, as well as many inorganic salts and gases. Its melting point is 18.5 °C (65.3 °F), and its boiling point is 189 °C (372.2 °F). It has a distinctive property of penetrating the skin very readily. Its taste has been described as oyster- or garlic-like. Other reported side effects include stomach upset, sensitivity to light, visual disturbances, and headache. Skin irritation can develop at the site where dimethyl sulfoxide is applied topically. Loading levels of 50–60 wt% are often observed compared with 10–20 wt% with typical solvents. For this reason, dimethyl sulfoxide plays a role in sample management and high-throughput screening operations in drug design. In cryobiology, dimethyl sulfoxide has been used as a cryoprotectant and it is still an important cryoprotectant for vitrification used to preserve organs, tissues, and cell suspensions.

Ethylene Glycol

Ethylene glycol, also known as monoethylene glycol or 1,2-ethanediol, is an alcohol with two hydroxyl groups (a diol). It may also be used as a protecting group for carbonyl groups in organic synthesis and it is widely used as an antifreeze in vehicles. In its pure form, it is an odorless, colorless, syrupy, sweet-tasting, toxic liquid. Its melting point is −12.9 °C (8.8 °F), and its boiling point is 197.3 °C (387.1 °F). The major use of ethylene glycol is as a medium for convective heat transfer in, for example, automobiles and liquid-cooled computers. Because of its low freezing point, it is used as a de-icing fluid for windshields and aircraft. It is also commonly used in chilled water air conditioning systems that place either the chiller or the air handler outside, or systems that must cool below the freezing temperature of water. Ethylene glycol is also used in the manufacture of some vaccines, but it is not itself present in the vaccines. The major toxicity from ethylene glycol is through ingestion, where it is oxidized to glycolic acid and then oxalic acid, which is toxic. Ethylene glycol and its toxic byproducts first affect the central nervous system, then the heart, and finally the kidneys. Ingestion of sufficient amounts can be fatal. Ethylene glycol is used widely for vitrification, especially oocyte and embryo vitrification.

Propylene Glycol

Propylene glycol, known also by its systematic name 1,2- propanediol and as 1,2-dihydroxypropane, methylethyl glycol, methylethylene glycol, Sirlene or Dowfrost, is an organic compound (a diol alcohol). Propylene glycol is usually a faintly sweet and colorless, clear viscous liquid that is hygroscopic and miscible with water, acetone, and chloroform. It contains an asymmetrical carbon atom, so it exists in two stereoisomers. Propylene glycol has properties similar to those of ethylene glycol. Pure optical isomers can be obtained by hydration of optically pure propylene oxide. Its melting point is −59 °C (−74.2 °F), and its boiling point is 188.2 °C (370.8 °F). Propylene glycol usually is used in antifreeze solutions, in hydraulic fluids, and as a solvent. It has numerous applications, for example as a solvent in many pharmaceuticals and as a less-toxic antifreeze, especially for human embryo cryopreservation.

Toxicity of Cryoprotectants

The toxicity of cryoprotectants refers to at least two effects. The first is the chemical reacting with cells before cryopreservation, and the second is the chemical causing the change of osmosis of freezing solutions. Relatively low concentrations of cryoprotectants are usually used in cryobiology and, therefore, the chemicals themselves may not be a major concern for toxicity, although the concentration of cryoprotectants used in rapid cooling is relatively high. For assessment of the toxicity of cryoprotectants, it seems necessary to consider the colligative property of the aqueous (freezing) solution, which may be related directly to the cell permeability of each cryoprotectant and may cause osmotic stress in the cells before freezing and thawing procedures. The permeating speed of cryoprotectants is related directly to temperature. Consequently, major factors to be considered in assessing the toxicity of cryoprotectants are their concentration, the exposure temperature, and the time in aqueous (freezing) solution.

Cryoprotectants can interact with each other in a mixture, or with crucial cell molecules, thereby producing effects other than those that would occur with an individual cryoprotectant. It has been observed that the protective effect of combinations of cryoprotectants can be greater than would be expected if the action of each agent were simply additive. Whether the toxicity of cryoprotectants can be reduced, or not, by mixing two or more cryoprotectants (in a system where there is a reduction in the concentration of each cryoprotectant) in the aqueous (freezing) solution needs to be further investigated.

Equilibration

It is common practice to suspend cells in aqueous (freezing) solution containing permeable cryoprotectants for the time that is required to equilibrate intracellular solutes before freezing. Many cells, especially eukaryotic cells, are sensitive to osmotic stress. Therefore, the permeating cryoprotectants are added gradually to the freezing solution in order to minimize osmotic stress as well as to allow removal of the cryoprotectants from the suspensions gradually. It means that the cells need time to balance and to adapt to osmotic shock. Different cell types require different equilibration times. Based on the permeating speed of the cryoprotectants and the cell types, different equilibration times will be required. Normally, lower temperature requires a longer period of equilibration, and higher temperature needs a shorter period.

The success of cryopreservation is determined by whether or not the cell undergoes intracellular ice formation during freezing. As mentioned above, vitrification can occur by either a slow-freezing or a rapidcooling procedure. In the slow freezing procedure, intracellular ice formation is avoided by sufficiently slow cooling that osmotic dehydration results in the water remaining in near chemical potential equilibrium with the outside solution and ice. In the rapid cooling procedure, the cooling rate needed for vitrification is approaching T_g. Different freezing solutions with different concentrations of cryoprotectants need different cooling rates for vitrification. These rates are derived primarily from calculations using mathematic al models of ice crystallization in very dilute solutions. However, the actual situation may differ from the theoretical rate in the different cell types.

Processes in Cryobiology

CHAPTER 2

Cryobiology is a vast subject that includes numerous processes such as cryosurgery, cryoablation, cryoimmunotheraphy, cryoneurolysis, cryotherapy, cryosurgery, cryo-transport, etc. The topics elaborated in this chapter will help in gaining a better perspective about various processes of cryobiology.

CRYOSURGERY

Cryosurgery involves tissue destruction under controlled freezing. The main advantages of this technique are that it is less invasive and has lower morbidity compared with surgical resection. However, the use of cryosurgery has been limited by a lack of good understanding of the underlying mechanisms of tissue destruction. The possible mechanisms of tissue injury are depicted in the figure below.

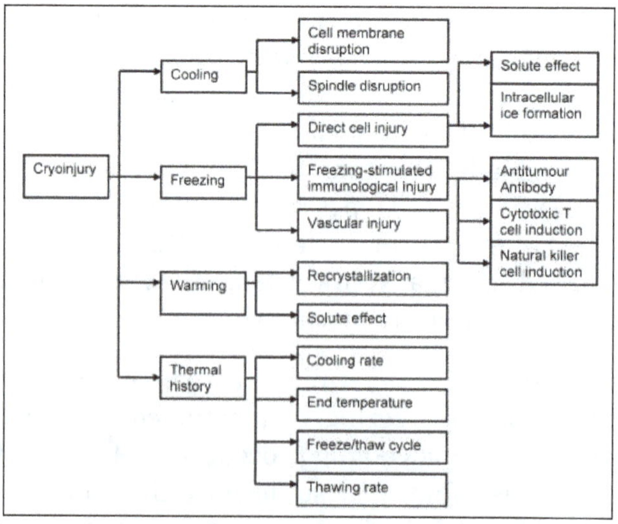

Flow chart showing the mechanisms of cryoinjury.

Each cell in tissue is exposed to different thermal histories. The cells nearest to the cryosurgical probe experience the lowest temperature and the fastest cooling rates compared with cells farther away from the probe. The temperature decreases until the heat extracted from the tissue equals the heat in the cooling fluid. Low cryogenic temperature is observed at the tissue surface in contact with the probe; near-body temperature is observed on the outer edge of lesion. There is then a period of thawing during which the probe is removed. Because different cells at different locations may experience different cooling temperatures for various periods of time, thermal history is important in assessing tissue damage.

Effects of Cooling

In general, most mammalian cells can withstand low, nonfreezing temperatures. However, these conditions can affect several aspects of cell function. The cell membrane is a lipid bilayer structure with proteins spanning across it. The cell membrane, in general, is impermeable except where membrane proteins allow mass transfer to occur. At low temperatures, the lipid transforms into a gel phase, or a structure with low free energy. During this process, the membrane proteins become separated and lose their ability to control mass transfer. The membrane becomes more permeable and allows ions to transfer in and out of cells more easily. As a result, the ionic composition of the cells changes and damage occurs.

The cytoskeleton is also affected by the cooling process. Meiotic spindles are known to be sensitive to hypothermia, resulting in tubulin depolymerization. In a study using human oocytes cooled to 0 °C for 2 min to 3 min, the meiotic spindle shortened and disappeared after 10 min.

Effects of Freezing

Extensive studies have been performed to understand the effect of freezing on biological tissue. One hypothesis is direct cellular injury from the extracellular space. The cell injury that occurs after freezing is thought to result from a high solute concentration causing cell dehydration. Intracellular ice formation causing intracellular organelle and cell membrane disruption has also been implicated.

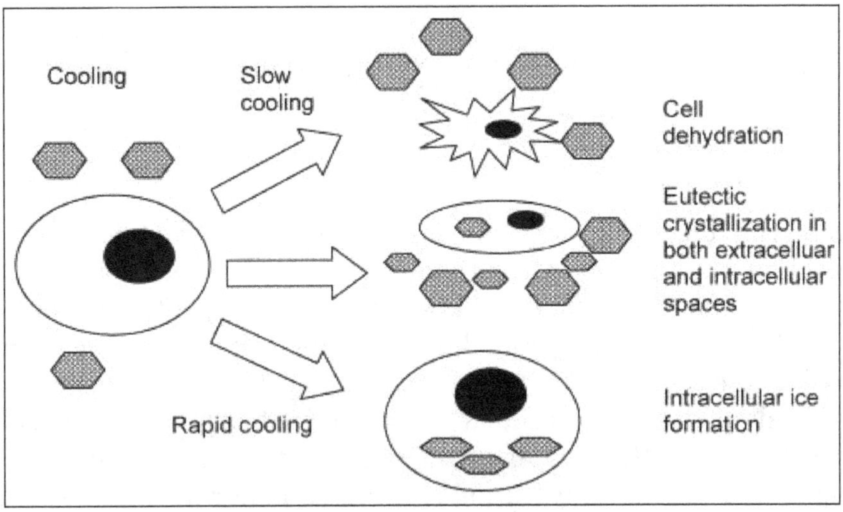

Diagram of the mechanisms of freezing injury.

During slow cooling, ice forms in the extracellular space because of the elevated solute concentration in the unfrozen fraction. This leads to cell shrinkage. When the temperature is further decreased, allowing the initiation of eutectic crystallization in the extracellular space, the temperature and concentration of the intracellular space may allow eutectic crystallization to occur in the intracellular space. Alternatively, if the cooling rate is rapid, the cells are not able to lose water fast enough to maintain equilibrium, resulting in intracellular water becoming supercooled and eventually frozen. Hexagons represent ice crystals.

Another theory is that freezing may stimulate immunological injury. It is believed that the immune system becomes sensitized to the destroyed frozen tissue, and any tissue left behind is attacked by the host's own immune system after cryosurgery. However, the relevance of immunological injury is still controversial. Finally, it has been theorized that freezing involves vascular injury. The hypothesis is that freezing results in stasis of blood flow, particularly in the capillaries. The resulting ischemia leads to tissue necrosis.

It is important to bear in mind that, unlike cryopreservation, where cells are frozen in vitro under uniform conditions and then stored frozen for long periods of time, tissue subjected to cryosurgery is frozen in vivo and its cells experience a wide temperature gradient.

Effects of Thawing

The effects of thawing depend on the previous cooling rate. It is known that slow thawing allows solute effects and maximum ice growth during recrystallization to take place. Because solute effects and ice growth are deleterious to cells, complete thawing before the start of another cycle is important in determining the success of cryosurgery for oncological conditions. However, rapid cooling followed by rapid thawing can also be beneficial. During rapid cooling, the ice crystals tend to be small, with high surface energies. The longer the time of thawing, the easier it is for the ice to recrystallize, especially for crystals with high surface energies. The larger ice crystals can be more destructive than the smaller ones, because of the size of the crystals or the forces generated during recrystallization. This is shown by the fact that red blood cells tend to have higher survival rates when cooled and warmed faster, or vice versa.

Effects of Thermal History

Other parameters of thermal history, including cooling rate, end temperature and hold time, are also important in modulating the degree of cell injury. Traditional cryosurgical techniques include rapid cooling, slow thawing and repetition of the freeze/thaw cycle. In general, rapid cooling is more destructive than slow cooling, and the response and threshold of the cooling rate are cell-dependent. Under the same freezing rate, tumour cells tend to retain more cellular water and are less susceptible to dehydration than normal cells. Also, differences in sensitivity exist for different malignant cells. For example, a greater degree of dehydration is observed in normal liver cells, followed by metastatic colon carcinoma and primary hepatocellular carcinoma.

Apart from the freezing rate, the lethality of freezing increases with decreasing end temperature. In a study using Dunning AT-1 rat prostate tumour cells, exposed to the same cooling rate as normal cells, cell viability decreased as a function of the end temperature and became zero at −60 °C. The sensitivity to end temperature is also cell-dependent. For instance, melanocytes and muscle cells are more sensitive to freezing, while other tumour cells, in general, are more resistant.

The hold time – the duration of time the tissue is in the frozen state – is also an important factor in determining the degree of cryoinjury. Cooling and holding human prostate cancer cells at −10 °C for 20 min had a more destructive pattern than holding for 10 min. In a study of different thermal parameters on canine skin, keeping the tissue frozen had a progressive destructive effect on cells.

Cell Injury during Cryosurgery

Direct Cell Injury

Direct cellular injury may occur through two different mechanisms. The first is the 'minimum volume' hypothesis. Under sufficiently slow cooling rates, freezing occurs in the extracellular space and the cells attempt to maintain equilibrium with the extracellular solution by osmosis. When cells are exposed to the high extracellular solute concentration during slow cooling, the cells shrink. As the solute concentration rises beyond a cell's limit to shrink, the chemical potential gradient is decreased by extracellular salt transfer to the cytoplasm, resulting in a high intracellular solute concentration. During the thawing process, the cell contents are highly concentrated compared with the extracellular solution. Additionally, sudden exposure to the hypotonic solution leads to osmotic rupture of cells. There are a number of caveats with regard to this concept of 'minimum volume'. One issue is that there is no actual minimum volume when precise measurements can be made. Second, it has been observed that the same degree of hemolysis can occur at different cell volumes using different concentrations of glycerol. Thus, while cell shrinkage and re-expansion are significant causes of cell damage, reduction of cell volume is probably not the predominant cause of cell injury.

The second putative mechanism of direct cell injury is membrane destabilization during freezing and thawing. Two different forms of injuries in cold nonacclimated protoplasts have been reported during freeze-induced dehydration. The first phenomenon occurs from 0 °C to −5 °C, when the protoplasts shrink to a minimum volume with almost 80% of cell water removed. During thawing, the are cells re-expanded but lysed before regaining their original volume. When the cells were initially cooled to lower temperatures, ie, at −10 °C, approximately 90% of the cell water was removed osmotically. However, when the cells thawed, they were osmotically unresponsive and did not re-expand. The investigators proposed that the membrane was damaged during dehydration, leading to the failure of water and solute molecules to re-enter the membrane during osmotic re-expansion.

Intracellular Ice Formation

Supercooling is defined as the process by which liquid maintains its liquid state below its freezing point. Supercooled water in a cell has a higher chemical potential and higher driving force to leave the cell and freeze externally. Therefore, when the cooling rate is slow, the cell becomes dehydrated. In contrast, if the cooling rate is sufficiently high, intracellular ice forms. The mechanism by which this process occurs remains controversial. The protein-pore theory proposes that extracellular ice propagates to the supercooled cytoplasm through the aqueous pore of the cell membrane. Ice growth during thawing is thought to be the cause of cell injury. This theory is supported by experiments in the salivary gland and confluent cell monolayers which demonstrated the propagation of ice via gap junctions with strong temperature dependence.

The surface-catalyzed theory hypothesizes that the interaction between extracellular ice and the plasma membrane, characterized by the contact angle between the cell membrane and ice, leads to the formation of intracellular ice. It is also posited that intracellular particles inside the cell can catalyze intracellular ice formation below −30 °C (volume-catalyzed nucleation).

The final theory is the membrane disruption theory. This proposes that intracellular ice formation occurs as a result of membrane disruption at the critical osmotic pressure gradient across the membrane during freezing.

While the exact mechanism of intracellular ice formation has still yet to be resolved, most cryobiologists agree that intracellular ice formation is lethal to cells. It is interesting, however, to note that on confluent monolayers of V-79W Chinese hamster fibroblasts and Madin-Darby canine kidney cells exposed to a slow cooling rate to temperatures as low as −40 °C, intracellular ice formation may confer cryoprotective effects from dehydration. Thus, the role and significance of intracellular ice formation has yet to be defined.

Mechanisms of Vascular Injury

A correlation between vascular injury and freezing was first proposed by Cohnheim in 1877, when he hypothesized that necrosis in frostbitten tissue was caused by stasis of blood flow after thawing. Later, Lewis and Love observed that vasculature in frozen human skin changed from stasis surrounded by an area of hyperemia at −5 °C to the previous normal state with surrounding edema when rewarmed to room temperature. Subsequent investigators have also shown that changes to the vasculature after freezing and thawing, such as increased tissue edema, circulatory stasis and progressive thrombosis, lead to tissue necrosis after frostbite. Taken together, these results support the hypothesis that vascular injury plays an important role in tissue injury.

Endothelial Damage in Vascular Injury

Extensive studies have been performed to investigate the role of the endothelium in mediating cryoinjury. Marzella et al studied freezing injury to rabbit ears by microscopy. They showed that the microvasculature endothelium was destroyed within 1 h. Platelet aggregation was observed immediately on thawing. Interstitial swelling and neutrophil recruitment also occurred minutes after thawing, with extravasation of red blood cells by 6 h and endothelial separation by 24 h.

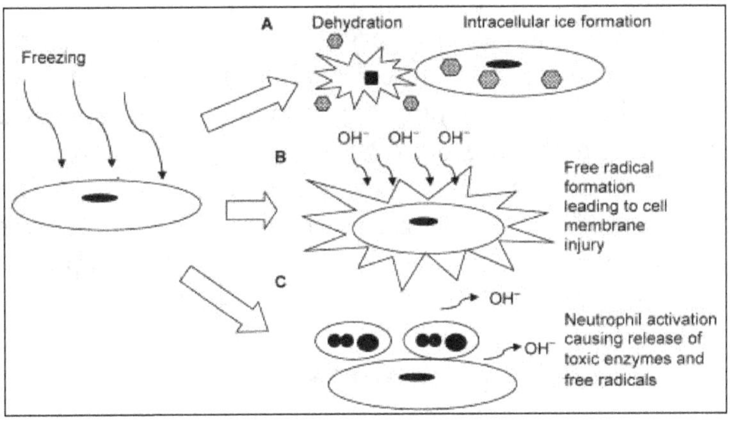

In figure above of mechanisms of endothelial injury from freezing. (A) Direct cell injury from dehydration in slow cooling conditions or intracellular ice formation in rapid cooling conditions. (B) Free radical production from lipid peroxidation, reduction of the electron transport chain of the inner mitochondrial membrane, or metabolism of hypoxanthine via the xanthine oxidase pathway. (C) Neutrophil activation, together with the production of free radicals and toxic enzymes, leads to cell membrane injury.

Several mechanisms have been proposed to explain endothelial injury after freezing. The first theory is direct cell injury. The second theory deals with free radical production. In a study of the role of free radicals after freezing and thawing, the administration of superoxide dismutase and deferoxamine improved the viability of rabbit ears after frostbite. Electron microscopy demonstrated that endothelial injury, vascular stasis, neutrophil adhesion and erythrocyte aggregation were present. Several mechanisms were proposed to explain the free radical formation during ischemia and thawing. The first mechanism postulates that the electron transport chain of the inner mitochondrial membrane may be reduced during ischemia, leading to oxygen radical formation. The second possible source of free radicals is from lipid peroxidation. During ischemia, there is an increase in free fatty acid and arachidonic acid. On thawing, blood flow returns and the accumulated arachidonic acid is metabolized via the lipoxygenase and cyclooxygenase pathways, leading to increased formation of thromboxanes and superoxides. Another proposed mechanism is the production of oxygen radicals by the metabolism of hypoxanthine, via the xanthine oxidase pathway during thawing.

Endothelial cells are the main source of xanthine oxidase in the blood vessels, implicating the importance of the interaction of endothelial cells with free radicals during freezing and thawing. The free radical theory is controversial because there are reports that fail to support this conclusion.

Another mechanism of post-thaw injury is by neutrophil activation. It is hypothesized that leukocytes may be trapped in the microvasculature, leading to obstruction. Leukocytes can also interact with the platelets and generate free radicals. The enzymes released by neutrophils damage the endothelial cells and increase the permeability of the endothelial cell layer via the production of active oxygen species. Gazzaniga et al studied the inflammatory changes after cryosurgery using human melanoma xenografted into nude mice. They found that endothelial cell activation was the first noticeable event, followed by infiltration of polymorphonuclear cells, and then macrophages. In contrast, using an intravital muscle model to study the microcirculatory changes in frostbite injury of rat muscle, other investigators found no significant role of neutrophil adhesion in the early response. Likewise, other groups reported that a similar area of tissue destruction was achieved whether the vascular supply was clamped or unclamped in hepatic cryosurgery.

Mechanisms of Immunological Injury

Several mechanisms of immunological injury have been proposed. The first theory is the production of antitumour antibodies. When the tumour cells die, the antigens inside the cells are released onto the membrane and phagocytosed by antigen-presenting cells. B cells with antibodies specific for the antigen are stimulated and transformed into plasma cells. Antibody formation induces complement fixation, leading to neutrophil and macrophage chemotaxis. These cells release free radicals and enzymes, which kill tumour cells left behind. This relationship of antibody production and cryosurgery is still controversial. Riera et al found that the antibody level decreased in isoimmunized rabbits following cryosurgery.

The second mechanism of immunological involvement is through the induction of cytotoxic T cells. Normally, intracellular antigens are transferred to the cell membrane and recognized by cytotoxic T cells, which release enzymes and kill the cells. It was proposed that cryosurgery may sensitize the cytotoxic T cells or change the antigen presentation. In a study performed by Eskandari et al, T cell activation peaked at two weeks after cryosurgery in a R3327 tumour in the Copenhagen rat, and remained elevated compared with the control group.

The third possible mechanism is that cryosurgery may stimulate the activity of natural killer cells. However, the relationship of cryosurgery and the activity of natural killer cells is still undetermined. Nevertheless, the response of the immune system to cryosurgery seems to be cell type-dependent. For instance, a positive response was demonstrated in R3327 prostate adenocarcinoma, but no effect was observed in MRMT-1 mammary adenocarcinoma. It may also be that the amount of antigen is important in immune system stimulation. If the antigen amount is more than the immune system can bear, suppression of tumour immunity may occur. Roy et al. found that the survival time decreased if a greater amount of cryodestroyed tumour was injected into the animal.

Cryosurgery for Common Skin Conditions

Cryosurgery is a highly effective treatment for a broad range of benign skin problems. Cryosurgery is best suited for use in patients with light skin and for treatment of lesions in most non–hair-bearing areas of the body. Spray methods include the timed spot freeze technique, the rotary or spiral pattern, and the paintbrush method. Benign skin lesions that are suitable for freezing include actinic keratosis, solar lentigo, seborrheic keratosis, viral wart, molluscum contagiosum, and dermatofibroma. Cryosurgery requires little time and fits easily into the physician's office schedule. Advantages of this treatment include a short preparation time, low risk of infection, and minimal wound care. In addition, cryosurgery requires no expensive supplies or injectable anesthesia, and the patient does not have to return for suture removal. Potential side effects include bleeding, blister formation, headache, hair loss, and hypopigmentation, but rarely scarring. Skin lesions often can be treated in a single session, although some require several treatments.

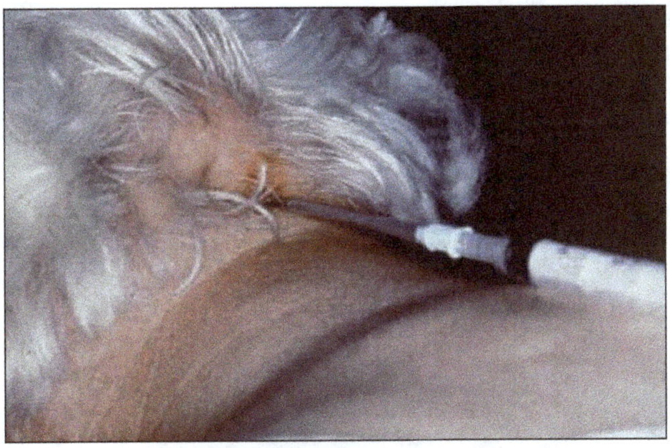

Cryosurgery has been used to treat skin lesions for approximately 100 years. The first cryogens were liquid air and compressed carbon dioxide snow. Liquid nitrogen became available in the 1940s and currently is the most widely used cryogen.

Over the past 50 years, much experience has been gained in the use of cryosurgery to treat skin lesions. The cotton-tipped dipstick method of liquid nitrogen application has been popular in the management of common benign lesions. However, this method is being supplanted by liquid nitrogen spray techniques. Liquid nitrogen spray equipment is easy to use, and similar techniques can be employed to manage benign, premalignant, and malignant lesions.

Cryosurgery devices. (Left) Cotton-tip applicator. (Center) Liquid nitrogen spray. (Right) Cryoprobe.

Mechanism of Action

Liquid nitrogen boils at −196 °C (−320.8 °F), is the most effective cryogen for clinical use. It is particularly useful in the treatment of malignant lesions. Temperatures of −25 °C to −50 °C (−13 °F to −58 °F) can be achieved within 30 seconds if a sufficient amount of liquid nitrogen is applied by spray or probe. Generally, destruction of benign lesions requires temperatures of −20 °C to −30 °C (−4 °F to −22 °F). Effective removal of malignant tissue often requires temperatures of −40 °C (−40 °F) to −50 °C.

Irreversible damage in treated tissue occurs because of intracellular ice formation. The degree of damage depends on the rate of cooling and the minimum temperature achieved. Inflammation develops during the 24 hours after treatment, further contributing to destruction of the lesion through immunologically mediated mechanisms.

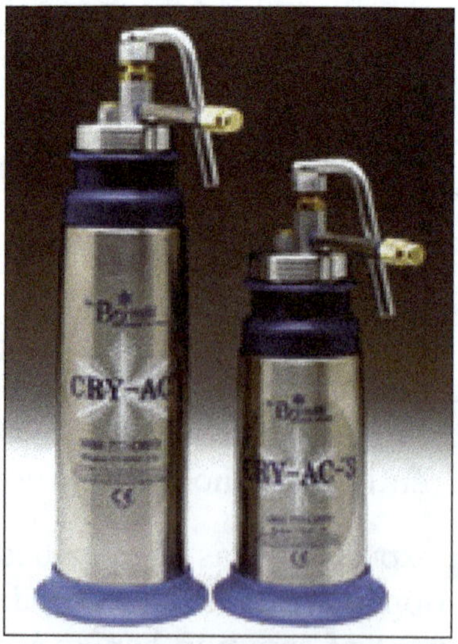

Liquid nitrogen spray equipment. (Left) Cryogun. (Right) Mini Cryogun.

Slow thaw times and repeat freeze-thaw cycles produce more tissue injury than a single freeze and thaw. Usually, several minutes are allowed between repeat freeze-thaw cycles. Repeat freeze-thaw cycles generally are employed only in the treatment of malignancy.

Continuous freezing at one location for more than 30 seconds beyond when an adequate freeze ball is achieved around the target area can result in disruption of the collagen matrix of the skin and possible scarring.

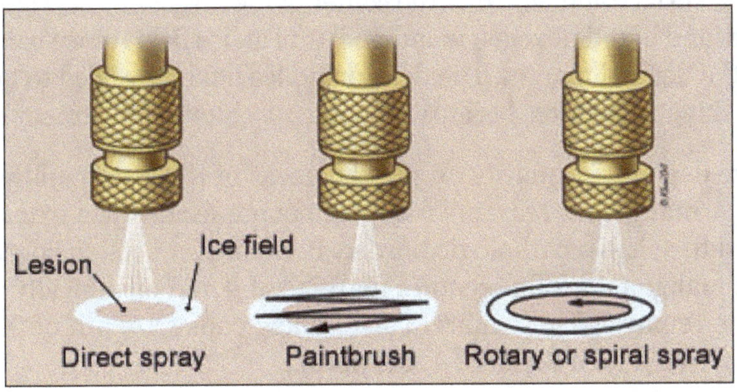

Liquid nitrogen spray patterns.

Mild freezing leads to dermoepidermal separation, which is useful in treating benign epidermal lesions. The more sensitive cells in the epidermis are destroyed while the dermis is left intact. Treatment may be complicated by an element of hypopigmentation, but studies and clinical experience indicate that repigmentation often occurs over several months because of undamaged melanocytes within hair follicles or the migration of melanocytes from the edge of the frozen zone. However, the predictability of repigmentation in individual patients is uncertain.

Methods of Treatment

The dose of liquid nitrogen and the choice of delivery method depend on the size, tissue type, and depth of the lesion. The area of the body on which the lesion is located and the required depth of freeze also should be considered. Additional patient factors to consider include the thickness of the epidermis and underlying structures, the water content of the skin, and local blood flow.

Liquid nitrogen spray methods for lesions of different sizes include the timed spot freeze or direct spray technique, the rotary or spiral pattern, and the paintbrush method.

Timed Spot Freeze Technique

The timed spot freeze technique allows greater standardization of liquid nitrogen delivery. It may be the most appropriate method for physicians who are learning to perform cryosurgery. Use of this technique maximizes the ability to destroy a lesion with minimal morbidity. The freezing time is adjusted according to variables such as skin thickness, vascularity, tissue type, and lesion characteristics.

Timed spot freezing is performed with a small spray gun that typically holds 300 to 500 mL of liquid nitrogen. Nozzle sizes range from A through F, with F representing the smallest aperture. Nozzle sizes B and C are suitable for the treatment of most benign and malignant lesions; they are the apertures most frequently noted in case reports.

For the standard spot freeze technique, the nozzle of the spray gun is positioned 1 to 1.5 cm from the skin surface and aimed at the center of the target lesion. The spray gun trigger is depressed, and liquid nitrogen is sprayed until an ice field (or ice ball) encompasses the lesion and the desired margin. The designated ice field may need to be delineated in advance with a skin marker pen, because freezing may blur pretreatment lesion margins.

The margin size depends primarily on the thickness of the lesion and whether the lesion is benign or malignant. Margins for most benign lesions can extend as little as 1 to 2 mm beyond the visible pathologic border. Premalignant lesions need margins of 2 to 3 mm, while malignant lesions require margins of 5 mm of clinically normal skin to ensure adequate removal. These margin sizes allow enough depth of freeze to ensure temperatures of −50 °C to a depth of 4 to 5 mm.

Once the ice field has filled the specified margin, the spray needs to be maintained, with the spray canister trigger pressure and, thus, the liquid nitrogen spray flow adjusted to keep the target field frozen for an adequate time. This time may vary from five to 30 seconds beyond the initial time for formation of the ice field. If more than one freeze-thaw cycle is required for lesion destruction, complete thawing should be allowed before the next cycle (usually two to three minutes).

The timed spot freeze technique achieves temperatures that are adequate for tissue destruction in an ice field up to 2 cm in diameter. The best approach for lesions larger than 2 cm (including an adequate margin) is to use overlapping treatment fields.

Other Techniques

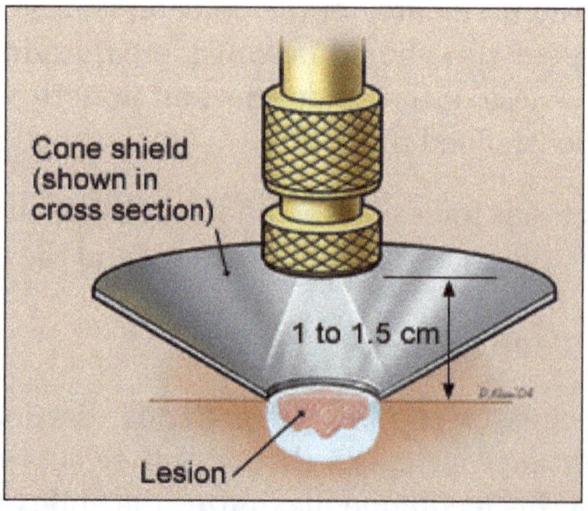

Liquid nitrogen spray application using a timed spot freeze technique and an open cone shield to direct the liquid nitrogen (note that an otoscope tip can be used with similar effect). The spray nozzle is positioned approximately 1 to 1.5 cm above the target lesion.

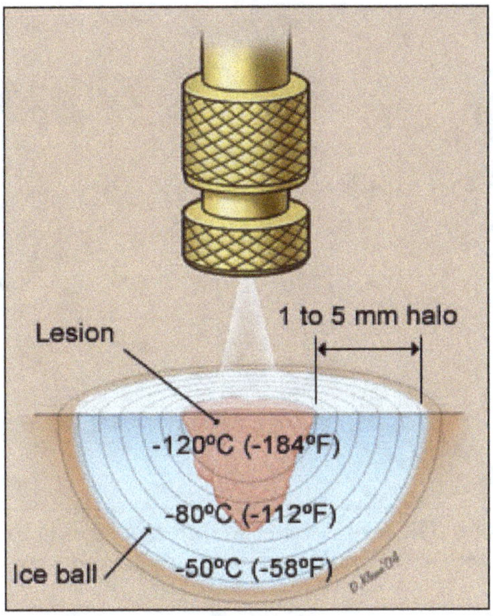

Timed spot freeze technique used to treat a malignancy (possibly a small basal cell cancer), demonstrating freeze ball formation and the 5-mm treatment margins necessary to achieve a temperature of −50°C (−58 °F) and, thus, the required depth of 4 to 5 mm.

Variations on the open spray technique include the rotary or spiral pattern and the paintbrush method. These techniques can be useful for treating larger benign lesions. They are not well standardized for ensuring the temperatures that are required for the destruction of malignant lesions.

Cryoprobes

While the open spray technique can be used for the most easily accessible lesions, a cryoprobe attached to the liquid nitrogen spray gun can provide added versatility, depending on the site and type of the lesion. Various sizes and types of cryoprobes are available. The cryoprobe is applied directly to the lesions. A gel interface medium often is used between the probe and the skin surface.

Cryoprobes frequently are used in the treatment of smaller facial lesions (e.g., on the eyelids), where scatter of liquid nitrogen is undesirable. Probes also are useful in the management of vascular lesions, where the pressure of the probe can be used to remove blood from the tissues thus allowing more adequate treatment.

Treatment of Benign Cutaneous Lesions

Most benign skin lesions can be treated successfully with any of several treatment modalities (excision, cryosurgery, electrodesiccation curettage). However, cosmesis, cost, and patient convenience may make one treatment modality more desirable than another. Patients should be informed about all treatment options and should be allowed to choose from the reasonable alternatives.

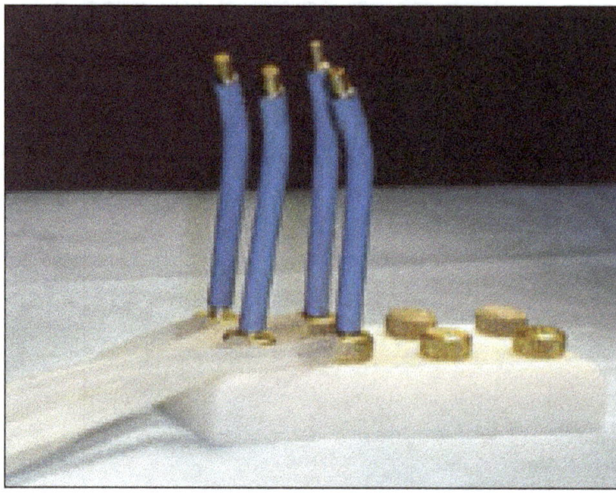

Mini Probe. The probe comes in different sizes (1 to 6 mm), with spray nozzles of variable aperture size.

Cryosurgery has advantages over the other modalities. Preparation time is short, and treatment requires no expensive supplies or injectable anesthesia. In addition, the risk of infection is low, wound care is minimal, and suture removal is not needed.

Correct clinical diagnosis and lesion selection are as critical as the timing of the liquid nitrogen spray in producing a favorable outcome. Depending on the type of lesion, standard technique may need to be varied. The table below summarizes cryosurgery techniques for a variety of benign skin lesions.

Sun-damaged Skin

Sun-damaged skin and related lesions are highly amenable to intervention with cryosurgical treatment. Localized small actinic keratosis, which is one of the most common solar-related skin abnormalities, usually requires only five to seven seconds of freeze time beyond initial appearance of a halo around the target lesion. Treatment requires only one freeze-thaw cycle and a margin of less than 1 mm.

Table: Recommended cryosurgical techniques.

Type	Technique	Freeze Time (Seconds)*	Number Of FTCS	Margin (MM)	Number Of Treatment Sessions	Interval (Weeks)†
Actinic keratosis	OS	5	1	1	1	
Cherry angioma	P	10	1	< 1	1	
Common warts	OS	10	1	2	3	4
Cutaneous horn	OS	10 to 15	1	2	1	
Dermatofibroma	P/OS	20 to 30	1	2	2	8
Hypertrophic scar	OS/P	20	1	2	1	

Ingrown toenail‡	OS	20	1	2	2	8
Keloid	OS/P	20 to 30	1	2	3	8
Myxoid cyst	OS/P	20	1	< 1	1	
Oral mucocele	P	10	1	< 1	1	
Pyogenic granuloma	OS	15	1	< 1	1	
Sebaceous hyperplasia	P	10	1	< 1	3	4
Skin tags	F/OS	5	1	2	1	
Solar lentigo	OS	5	1	< 1	1	

Seborrheic Keratosis

Seborrheic keratosis, the most common benign neoplasm, is best treated with cryosurgery or shave excision/curettage. Cryosurgery is especially effective in patients with multiple lesions.[11] Thin, flat lesions usually require only one five- to 10-second freeze-thaw cycle; larger, thicker lesions may need longer treatments times or, occasionally, two freeze-thaw cycles.

In treating seborrheic keratosis, the physician should consider the potential for hair loss in treated areas when choosing a therapeutic modality. The other major side effect of cryosurgery for this lesion is hypopigmentation. This side effect is more likely to occur in patients with dark skin.

Viral Skin Infections

Warts that are resistant to over-the-counter topical agents commonly are treated with cryosurgery. However, response is variable and often depends on the size of the wart and the degree of hyperkeratosis. Several treatment sessions typically are required, and the overall success rate is approximately 75 percent.

To avoid hypopigmentation, small flat warts may be treated with a light spray technique. Digital warts respond favorably to the timed spot freeze spray technique. Deep plantar or palmar warts present challenges, because pain may limit the patient's tolerance of freezing. However, favorable cure rates have been reported for initial pretreatment with keratolytics (salicylic acid). The wart then can be shaved, and the base can be frozen with or without multiple freeze-thaw cycles.

Cryosurgery has been found to be effective in the management of condyloma acuminatum, particularly when treatment with podophyllin (Podocon-25) has failed or the lesion is located in an area where use of this agent is undesirable.

Molluscum contagiosum, a common dermatologic problem in younger persons, is caused by a poxvirus. This lesion is amenable to cryosurgery, if indicated. Applying liquid nitrogen spray for a few seconds until the surface of the umbilicated papule turns white usually is adequate.

Table: Contraindications to cryosurgery.

Absolute contraindications	Relative contraindications
Lesion for which tissue pathology is required	Cold intolerance
Lesion located in an area with compromised circulation	Cold urticarial
Melanoma	Collagen disease or autoimmune disease
Patient unable to accept possibility of pigmentary changes	Concurrent treatment with immunosuppressive drugs
Proven sensitivity or adverse reaction to cryosurgery	Cryoglobulinemia
Sclerosing basal cell carcinoma or recurrent basal cell or squamous cell carcinoma, particularly when located in a high-risk area (e.g., temple, nasolabial fold)	Heavily pigmented skin
	Lesions located in pretibial areas, eyelid margins, nasolabial fold, ala nasi, and hair-bearing areas
	Multiple myeloma
	Pyoderma gangrenosum
	Raynaud's disease

Dermatofibroma

Open spray or cryoprobe techniques may be used to improve the cosmetic appearance of dermatofibromas. Surgical excision of these deep, asymptomatic skin nodules may result in hypertrophic scar formation, because the fibrous lesions are thought to arise from skin microtrauma.

Table: Complications and side effects of cryosurgery.

Acute	Delayed	Protracted or permanent	Protracted but temporary
Bleeding at the freeze site	Bleeding	Atrophy (rare)	Alteration of sensation
Blister formation	Excess granulation tissue formation (rare)	Hair and hair follicle loss	Hyperpigmentation
Edema	Infection (rare)	Hypopigmentation	Hypertrophic scarring
Headache (after treatment of facial lesions)			Milia
Pain			
Syncope (vasovagal; rare)			

A single 20- to 30-second freeze-thaw cycle is advised, and a 1- to 2-mm margin should be obtained. Retreatment in eight weeks may be necessary. Significant clinical provement, including visible flattening of raised dermatofibromas and lightening of pigmentation, has been reported in 80 to 90 percent of patients.

Contraindications

The relatively few contraindications to cryosurgery generally are related to concomitant illnesses in which excess reactions to cold may occur or delayed healing may be anticipated. Some relative contraindications may make alternative treatment modalities more suitable. Physicians often do not perform cryosurgery in the pretibial areas, especially in elderly patients, because of slow wound healing.

Complications

Common complications and side effects of cryosurgery are listed in the table. Skin discomfort, generally a burning sensation, occurs with cryosurgery, but intensity is variable. The most sensitive areas are the fingertips, ears, and temples. Freezing of lesions on the forehead or temple may produce headaches. Treatment in hair-bearing areas can result in permanent hair loss.

Hypopigmentation is common, especially with longer freeze times, but is less noticeable in light-skinned patients and improves within several months. Hypopigmentation is caused by the greater sensitivity of melanocytes to freezing, a situation that can be used to advantage in the treatment of dermatofibromas, which frequently have some mild overlying hyperpigmentation. Feathering of the freeze margin (lighter freeze area) often results in a better transition of pigmentary changes.

Freezing for less than 30 seconds beyond initial freeze ball formation does not result in scarring because of the preservation of fibroblasts and the collagen layer of the dermis, which allows for in-migration of the cellular components in the healing process and normal integrity of the skin layers.

Although rare and usually temporary, sensory nerve damage has been reported occasionally in large case series. It may take 12 to 18 months for sensation to return.

Cryosurgery in Cancer Treatment

Cryosurgery is also used to treat tumors inside the body (internal tumors and tumors in the bone). For internal tumors, liquid nitrogen or argon gas is circulated through a hollow instrument called a cryoprobe, which is placed in contact with the tumor. The doctor uses ultrasound or MRI to guide the cryoprobe and monitor the freezing of the cells, thus limiting damage to nearby healthy tissue. (In ultrasound, sound waves are bounced off organs and other tissues to create a picture called a sonogram.) A ball of ice crystals forms around the probe, freezing nearby cells. Sometimes more than one probe is used to deliver the liquid nitrogen to various parts of the tumor. The probes may be put into the tumor during surgery or through the skin (percutaneously). After cryosurgery, the frozen tissue thaws and is either naturally absorbed by the body (for internal tumors), or it dissolves and forms a scab (for external tumors).

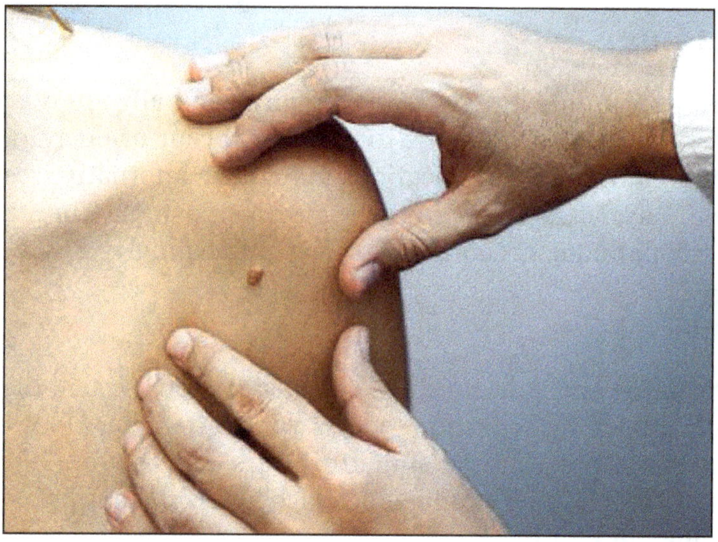

Cryosurgery is used to treat several types of cancer, and some precancerous or noncancerous conditions. In addition to prostate and liver tumors, cryosurgery can be an effective treatment for the following:

- Retinoblastoma (a childhood cancer that affects the retina of the eye). Doctors have found that cryosurgery is most effective when the tumor is small and only in certain parts of the retina.
- Early-stage skin cancers (both basal cell and squamous cell carcinomas).
- Precancerous skin growths known as actinic keratosis.
- Precancerous conditions of the cervix known as cervical intraepithelial neoplasia (abnormal cell changes in the cervix that can develop into cervical cancer).

Cryosurgery is also used to treat some types of low-grade cancerous and noncancerous tumors of the bone. It may reduce the risk of joint damage when compared with more extensive surgery, and help lessen the need for amputation. The treatment is also used to treat AIDS-related Kaposi sarcoma when the skin lesions are small and localized.

Researchers are evaluating cryosurgery as a treatment for a number of cancers, including breast, colon and kidney cancer. They are also exploring cryotherapy in combination with other cancer treatments, such as hormone therapy, chemotherapy, radiation therapy, or surgery.

Cryosurgery can be used to treat men who have early-stage prostate cancer that is confined to the prostate gland. It is less well established than standard prostatectomy and various types of radiation therapy. Long-term outcomes are not known. Because it is effective only in small areas, cryosurgery is not used to treat prostate cancer that has spread outside the gland, or to distant parts of the body.

Some advantages of cryosurgery are that the procedure can be repeated, and it can be used to treat men who cannot have surgery or radiation therapy because of their age or other medical problems.

Cryosurgery for the prostate gland can cause side effects. These side effects may occur more often in men who have had radiation to the prostate.

- Cryosurgery may obstruct urine flow or cause incontinence (lack of control over urine flow); often, these side effects are temporary.
- Many men become impotent (loss of sexual function).
- In some cases, the surgery has caused injury to the rectum.

Cryosurgery may be used to treat primary liver cancer that has not spread. It is used especially if surgery is not possible due to factors such as other medical conditions. The treatment also may be used for cancer that has spread to the liver from another site (such as the colon or rectum). In some cases, chemotherapy and/or radiation therapy may be given before or after cryosurgery. Cryosurgery in the liver may cause damage to the bile ducts and/or major blood vessels, which can lead to hemorrhage (heavy bleeding) or infection.

Side Effects of Cryosurgery

Cryosurgery does have side effects, although they may be less severe than those associated with surgery or radiation therapy. The effects depend on the location of the tumor. Cryosurgery for cervical intraepithelial neoplasia has not been shown to affect a woman's fertility, but it can cause cramping, pain, or bleeding. When used to treat skin cancer (including Kaposi sarcoma), cryosurgery may cause scarring and swelling; if nerves are damaged, loss of sensation may occur, and, rarely, it may cause a loss of pigmentation and loss of hair in the treated area. When used to treat tumors of the bone, cryosurgery may lead to the destruction of nearby bone tissue and result in fractures, but these effects may not be seen for some time after the initial treatment and can often be delayed with other treatments. In rare cases, cryosurgery may interact badly with certain types of chemotherapy. Although the side effects of cryosurgery may be less severe than those associated with conventional surgery or radiation, more studies are needed to determine the long-term effects.

Advantages of Cryosurgery

Cryosurgery offers advantages over other methods of cancer treatment. It is less invasive than surgery, involving only a small incision or insertion of the cryoprobe through the skin. Consequently, pain, bleeding, and other complications of surgery are minimized. Cryosurgery is less expensive than other treatments and requires shorter recovery time and a shorter hospital stay, or no hospital stay at all. Sometimes cryosurgery can be done using only local anesthesia.

Because physicians can focus cryosurgical treatment on a limited area, they can avoid the destruction of nearby healthy tissue. The treatment can be safely repeated and may be used along with standard treatments such as surgery, chemotherapy, hormone therapy, and radiation. Cryosurgery may offer an option for treating cancers that are considered inoperable or that do not respond to standard treatments. Furthermore, it can be used for patients who are not good candidates for conventional surgery because of their age or other medical conditions.

Disadvantages of Cryosurgery

The major disadvantage of cryosurgery is the uncertainty surrounding its long-term effectiveness. While cryosurgery may be effective in treating tumors the physician can see by using imaging tests (tests that produce pictures of areas inside the body), it can miss microscopic cancer spread. Furthermore, because the effectiveness of the technique is still being assessed, insurance coverage issues may arise.

CRYOABLATION

Cryoablation is a process that uses extreme cold to destroy tissue. Cryoablation is performed using hollow needles (cryoprobes) through which cooled, thermally conductive, fluids are circulated. Cryoprobes are positioned adjacent to the target in such a way that the freezing process will destroy the diseased tissue. Once the probes are in place, the attached cryogenic freezing unit removes heat from ("cools") the tip of the probe and by extension from the surrounding tissues.

Ablation occurs in tissue that has been frozen by at least three mechanisms:

- Formation of ice crystals within cells thereby disrupting membranes, and interrupting cellular metabolism among other processes,
- Coagulation of blood thereby interrupting bloodflow to the tissue in turn causing ischemia and cell death,
- Induction of apoptosis, the so-called programmed cell death cascade.

The most common application of cryoablation is to ablate solid tumors found in the lung, liver, breast, kidney and prostate. The use in prostate and renal cryoablation are the most common. Although sometimes applied in cryosurgery through laparoscopic or open surgical approaches, most often cryoablation is performed percutaneously (through the skin and into the target tissue containing the tumor) by a medical specialist, such as an interventional radiologist.

Prostate

Prostate cryoablation is moderately effective but, as with any prostate removal process, also can result in impotence. Prostate cryoablation is used in three patient categories:

- As primary therapy in patients for whom sexual function is less important or who are poor candidates for radical retropubic prostatectomy (RRP, surgical removal of the prostate),
- As salvage therapy in patients who have failed brachytherapy (the use of implanted radioactive "seeds" placed within the prostate) or external beam radiation therapy (EBRT),
- Focal therapy for smaller, discrete tumors in younger patients.

Bone Cancer

Cryoablation has been explored as an alternative to radiofrequency ablation in the treatment of moderate to severe pain in people with metastatic bone disease. The area of tissue destruction created by this technique can be monitored more effectively by CT than RFA, a potential advantage when treating tumors adjacent to critical structures.

Renal

Cryoablation has similar outcomes to radiofrequency ablation when treating renal cell carcinoma.

Breast Cancer

Cryoablation for breast cancer is typically only possible for small tumors. Often surgery is used following cryoablation.

Cardiac

Another type of cryoablation is used to restore normal electrical conduction by freezing tissue or heart pathways that interfere with the normal distribution of the heart's electrical impulses. Cryoablation is used in two types of intervention for the treatment of arrhythmias: (1) catheter-based procedures and (2) surgical operations.

A catheter is a very thin tube that is inserted into a vein in the patient's leg and threaded to the heart where it delivers energy to treat the patient's arrhythmia. In surgical procedures, a flexible probe is used directly on an exposed heart to apply the energy that interrupts the arrhythmia. By cooling the tip of a cryoablation catheter (cardiology) or probe (heart surgery) to sub-zero temperatures, the cells in the heart responsible for conducting the arrhythmia are altered so that they no longer conduct electrical impulses.

Fibroadenoma

Cryoablation is also currently being used to treat fibroadenomas of the breast. Fibroadenomas are benign breast tumors that are found in approximately 10% of women (primarily ages 15–30).

In this procedure which has been approved by the U.S. Food and Drug Administration (FDA), an ultrasound-guided probe is inserted into the fibroadenoma and extremely cold temperatures are then used to destroy the abnormal cells. Over time the cells are reabsorbed into the body. The procedure can be performed in a doctor's office setting with local anesthesia and leaves very little scarring compared to open surgical procedures.

Catheter-based Procedures

Different catheter-based ablation techniques may be used and they generally fall into two categories: (1) cold-based procedures where tissue cooling is used to treat the arrhythmia, and (2) heat-based procedures where high temperature is used to alter the abnormal conductive tissue in the heart.

Cryoablation

Cold temperatures are used in cryoablation to chill or freeze cells that conduct abnormal heart rhythms. The catheter removes heat from the tissue to cool it to temperatures as low as -75 °C. This causes localized scarring, which cuts undesired conduction paths.

This is a much newer treatment for supraventricular tachycardia (SVT) involving the atrioventicular (AV) node directly. SVT involving the AV node is often a contraindication for using radiofrequency ablation because of the risk of injuring the AV node, forcing patients to receive a permanent pacemaker. With cryoablation, areas of tissue can be mapped by limited, reversible, freezing (e.g., to -10 C). If the result is undesirable, the tissue can be rewarmed without permanent damage. Otherwise, the tissue can be permanently ablated by freezing it to a lower temperature (e.g., -73 C).

This therapy has revolutionized AV nodal reentrant tachycardia (AVNRT) and other AV nodal tachyarrhythmias. It has allowed people who were otherwise not a candidate for radiofrequency ablation to have a chance at having their problem cured. This technology was developed at The Montreal Heart institute in the late 1990s. The therapy was successfully adopted in Europe in 2001, and in the USA in 2004 following the "Frosty Trial".

In 2004, the technology was pioneered in the midwest United States at Miami Valley Hospital in Dayton, Ohio by Mark Krebs, MD, FACC, Matthew Hoskins, RN, BSN and

Ken Peterman, RN, BSN. These electrophysiology experts were successful in curing the first 12 candidates in their facility.

Cryoablation for AVNRT and other arrhythmias do have some drawbacks. A recent study concluded that procedure times are slightly higher on average for cryoablation than for traditional radio-frequency (heat-based) ablations. Also, higher rate of equipment failures were recorded using this technique. Finally, even though short term success rate is equivalent to RF treatments, cryoablation appears to have a significantly higher long term recurrence rate.

Site Testing

Cryotherapy is able to produce a temporary electrical block by cooling down the tissue believed to be conducting the arrhythmia. This allows the physician to make sure this is the right site before permanently disabling it. The ability to test a site in this way is referred to as site testing or cryomapping.

When ablating tissue near the AV node (a special conduction center that carries electrical impulses from the atria to the ventricles), there is a risk of producing heart block – that is, normal conduction from the atria cannot be transmitted to the ventricles. Freezing tissue near the AV node is less likely to provoke irreversible heart block than ablating it with heat.

Surgical Procedures

As in catheter-based procedures, techniques using heating or cooling temperatures may be used to treat arrhythmias during heart surgery. Techniques also exist where incisions are used in the open heart to interrupt abnormal electrical conduction (Maze procedure). Cryosurgery involves the use of freezing techniques for the treatment of arrhythmias during surgery.

A physician may recommend cryosurgery being used during the course of heart surgery as a secondary procedure to treat any arrhythmia that was present or that may appear during the primary openchest procedure. The most common heart operations in which cryosurgery may be used in this way are mitral valve repairs and coronary artery bypass grafting. During the procedure, a flexible cryoprobe is placed on or around the heart and delivers cold energy that disables tissue responsible for conducting the arrhythmia.

Cryoablation and Immunotherapy

Cryoablation and the Immune System

Cryoablation may be synergistic with the immune system in the way cell death is induced. After ablation, the tumour remains, releasing various factors attracting the immune system. A proposed theory regarding the potential mechanisms of cryoablation and the immune system is the danger theory of Matzinger. This theory proposes that after cell death by necrosis, cells secrete danger signals. These danger signals can initiate an immune response. In addition, these signals can mature dendritic cells (DCs) to fully activate T cells which may lead to a specific immune response. Cryoablation induces cell death by necrosis in which intracellular contents are still preserved while DNA, RNA and heat shock protein (HSP), which can induce danger signals, are released. On the other hand, the cells in the outer margin of cryoablated tissue die from apoptosis and do not release DNA, RNA and HSP, and with no danger signal, the DCs remain immature. Immature DCs may trigger immune suppressive signals that could lead to anergy (T cell inactivation). Therefore, cryoablation can induce both an immunostimulatory and immunosuppressive response.

In addition, cytokines are produced after cryoablation, and these can also influence an immune response. Again, both immunosuppressive and immunostimulatory cytokines may be released depending on the tumour tissue, age and freeze rate. For cryoablation of liver tumours, when more than 20% of the liver volume is ablated, a systemic inflammatory response can occur due to a release of cytokines interleukine-6 (IL-6), IL-10 and tumour necrosis factor alpha (TNFα), which can have marked systemic effects. Therefore, cryoablation is not the preferred treatment since heat-based ablations are better established in this setting. Yet, two clinical studies reported favourable outcomes on overall survival when cryoablation was combined with immunotherapy of allogenic natural killer (NK) cell infusion and dendritic cell cytokine-induced killer (DC-CIK) cells.

Preclinical work confirms the effect of cryoablation on the immune system. This study reported that cryoablation creates an antigen depot resulting in maturation of DCs. The maturation of DCs led to a tumour-specific immune response protecting half of the mice against a new injection of similar tumour cells. When cryoablation was combined with an immune checkpoint inhibitor, anti-cytotoxic T lymphocyte-associated protein 4 (CTLA-4) antibody, this anti-tumour effect was further enhanced with up to 80% of the mice becoming tumour free.

Clinical work evaluated peripheral blood after heat- and cold-based ablation modalities with pre-treatment (baseline) in different tumours showing no change in T cell subtypes (regulatory T cell (Tregs), T1 helper or Th2 helper cells) in all modalities, an elevation of cytotoxic T cells after heat-based ablative treatment was noted, and this was not identified after cryoablation. In a group of hepatitis B-positive hepatocellular carcinoma patients, the presence of elevated programmed cell death protein 1 (PD-1) on T cells and programmed cell death ligand 1 (PD-L1) expression on tumour cells had a poor overall survival post cryoablation. Theoretically, when combining cryoablation with a PD-1 inhibitor, such as nivolumab or pembrolizumab, the cryoablation induced adaptive immune resistance with upregulation of PD-L1 on tumour cells could be overcome, resulting in an effective anti-tumour T cell response. This potential synergy between cryoablation and anti-PD-1 may result in a more effective disease control.

To enhance the immunogenicity effect of the cryoablation, different immunotherapies can be given. Most immunotherapy enhances the innate immunity, namely natural killer (NK) cell therapy, dendritic cells (DCs) and CpG oligonucleotide (CpG ODN). When cryoablation presents the contents of the tumour cells, the innate immune system can assimilate these contents and present it to T cells; an immune response specific to the tumour cells could be obtained.

Table: The major therapies with their mode of action used in combination with cryoablation. All therapies stimulate the immune system in a way and in combination with cryoablation an enhancement of this effect is hypothesized.

Therapy	Mode of action
CpG oligonucleotide (CpG ODN)	Is recognized by dendritic cells (DCs) and B cells. Activates T cells, natural killer (NK) cells, monocytes, neutrophils and plasma cell differentiation.
Anti-cytotoxic T lymphocyte-associated protein 4 (anti CTLA-4)	Blocks the inhibitory receptor (CTLA-4) on the T cell and therefore activates the T cell for a specific immune response.
Immature dendritic cells (DCs)	Phagocytosis of pathogens; antigen-presentation to other immune cells (among others T cells).
Natural killer (NK) cell therapy	Infusion with autologous NK cells to directly destroy tumour cells.
Dendritic cell- cytokine induced killer (DC-CIK)	May act similarly to T cells or NK cells but is unrestricted to major histocompatibility complex.
Granulocyte-macrophage colony-stimulating factor (GM-CSF)	A protein that functions as a cytokine and stimulates stem cells and can induce an immune cascade.
Anti-programmed death-ligand 1 (PDL-1)	Blocks the receptor programmed death 1 on the tumour cell. This results in the activation of the T cell to induce a specific immune response.

Breast Cancer

Cryoablation was approved for the treatment of fibroadenomas for over a decade. In 2016, the American College of Surgeons Oncology Group (ACOSOG) alliance considered cryoablation as an effective treatment for unifocal ductal cancer, with a success rate of complete tumour ablation of 92% after correction for multifocal disease. Also, for the treatment of stage IV breast cancer, cryoablation is a safe and effective procedure to control the disease and debulks the tumour in the breast.

Although antibodies are part of the treatment for human epidermal growth factor receptor 2 (HER2)-positive breast cancers, no active forms of immunotherapy such as immune checkpoint inhibitors are currently approved for breast cancer. Recently, the interim analysis of the Impassion 130 study with the combination of nab-paclitaxel plus atezolizumab revealed an impressive 10 months of improvement in overall survival compared with chemotherapy alone in PD-L1-positive triple negative metastatic breast cancers. Other immune checkpoint blocking antibodies, such anti-CTLA-4 antibody, are also under investigation.

One of the first reports confirming the immunogenicity of cryoablation in breast cancer was in a mammary mouse model. After cryoablation or surgery, mice were re-challenged with tumour cells and only 16% of the cryoablated exhibited tumour development compared to 86% of the surgically treated mice.

In a metastatic breast cancer mouse model, cryoablated mice treated with a high freeze rate (100% cryoablation cycle) showed an improvement in overall survival with significant reduction in the number of pulmonary metastasis compared to treatment with a low freeze rate (10% cryoablation cycle) or those treated with surgery. When cryoablation was combined with injection of CpG ODN (a single-strand DNA molecule that acts as a toll-like receptor (TLR) agonist to stimulate and mature DCs) in mice, less tumour recurrence and secondary tumour growth were seen after the re-challenge compared to the mice that received cryoablation alone or surgery. No significant difference was reported between CpG ODN alone or in combination with cryoablation groups, leaving the added role of cryoablation to CpG ODN injection undetermined regarding cytokine release and potential immune activation.

Other studies have investigated the combination of cryoablation with an immune checkpoint inhibitor (ipilimumab) 7 days before mastectomy in a group of patients with early-stage breast cancer. The pilot study, including 19 patients, showed that this approach was a safe option without delaying the mastectomy. A post hoc analysis was performed to assess the possibility of T cell receptor sequencing as a biomarker for T cell response to cryoablation, where no specific T cell response was observed. Additionally, a phase II trial is ongoing where cryoablation in combined with ipilimumab and nivolumab before breast surgery in triple negative breast cancer patients after taxane-based neoadjuvant chemotherapy.

In another study, recurrent HER2-positive breast cancer patients were treated with the combination of cryoablation, trastuzumab and natural killer (NK) cell therapy (intravenous infusion of allogenic NK cells). These patients displayed a significantly prolonged progression-free survival (PFS), significantly larger numbers of T cells and Th1 cytokines together with a significantly reduction in the number of circulation tumour cells in the peripheral blood compared to patients only treated with cryoablation alone or cryoablation and NK cell therapy. To note, PFS was not reached in the triple combination group, and the significantly prolonged PFS could be due to trastuzumab. Niu et al. evaluated the use of cryoablation in combination with immunotherapy of DC-CIK cells in metastatic breast cancer patients versus chemotherapy or cryoablation alone. The group of patients that received multiple cryoablations (several sites) in combination with immunotherapy displayed a significantly longer median overall survival compared to the other groups.

Together, cryoablation before mastectomy is feasible and its combination with immunotherapy, consisting of NK cell therapy, DC-CIK or anti-CTLA-4 antibody, is safe and effective in different stages in breast cancer. These results encourage further investigation into the combination of cryoablation and immunotherapy for breast cancer patients. Two trials are currently open which combine cryoablation with immune checkpoint inhibitors.

Renal Cell Carcinoma

In RCC, cryoablation is most frequently used to treat stage I cancer (ideally smaller than 4 cm taken as the largest diameter) in patients not eligible for surgical resection. With optimal patient selection, results similar to partial nephrectomy can be achieved.

Immunotherapy for RCC has been used for quite some time, and nivolumab, a PD-1 inhibitor, is already approved for the treatment of RCC.

Two animal studies showed the favourable effect of cryoablation in the microenvironment of RCC and in the kidney. The first study used two mice models, one with and one without injected RCC to observe an inflammatory immune response after cryoablation in the tumour or healthy kidney tissue. An infiltration of neutrophils, macrophages and CD4+ and CD8+ T cells was reported after cryoablation whereby no difference was observed after cryoablation of normal kidney tissue or tumour tissue. Another study compared cryoablation with surgery and showed decreased tumour growth after the re-challenge of the tumour cells with significantly more T cells in the peripheral blood after cryoablation.

Kato et al. showed that in half of the patients with T1 RCC, a significant increase in T cell receptor (TCR) B CD3 clonotypes of T-cells in post ablation tissue and blood was seen with a low diversity (TCR clones were not evenly distributed anymore). In another clinical study, two sessions of cryoablation of the pulmonary metastases, each combined with two Intratumoural injections of granulocyte-macrophage colony-stimulating factor (GM-CSF), resulted in higher levels of NK cells, Th1 cytokines and T and B cells in the peripheral blood compared to baseline. Lin et al. showed similar effects of allogeneic NK cell immunotherapy combined with cryoablation in 60 advanced RCC patients, and this treatment combination resulted in more tumour responses and decrease in Hounsfield units count than cryoablation on its own.

To summarise, cryoablation of RCC elicits an immune response and can be safely combined with GM-CSF and NK cell therapy. Currently, one trial is ongoing investigating the synergy of cryoablation with anti-PD-1 therapy (tremelimumab), and another trial investigates the effect of ablation of the immune system.

Prostate Cancer

Cryoablation is currently being used to treat stage I prostate cancer. Cryoablation could also be considered as salvage treatment for local recurrence after radiation therapy.

Future perspectives in prostate cancer shift towards a more targeted therapy where cryoablation may have an important role in prostate cancer.

Presently in prostate cancer, the only approved immunotherapy is sipuleucel-T (Provenge), a DC-based immunotherapy that sensitises dendritic cells with prostate antigens and is used as a therapeutic vaccine. Other immunotherapies evaluated have failed to show improvement in overall survival. Current developments are focusing on immunotherapy for the subgroup with defects in DNA-repair mechanisms, which include microsatellite instability and breast cancer gene mutations.

A regression in secondary tumour growth with infiltration of CD4+ and CD8+ T cells and lower counts of Tregs in mice treated by cryoablation and anti-CTLA-4 antibodies. Another study reported that the combination of cryoablation with anti-CTLA-4 antibodies reduced distant metastasis in mice together with a reduction in the number of Tregs; these were lowest on day 14 but returned to normal levels at day 21. Recently, combination therapy of androgen deprivation plus anti-PD-1, anti-CLTA-4 or placebo with or without cryoablation demonstrated a delay in distant tumour growth and decreased mortality in mice in the trimodal therapy groups.

Pre cryoablation tissue in high-risk Localised prostate cancer patients showed elevated numbers of Tregs compared to healthy volunteers. Numbers of Tregs decreased significantly after cryoablation in the prostate patients, and, conversely, 7 of 12 patients had an increase of suppressive function of Tregs measured by immunosuppressive assay of $CD4^{+}CD25^{+}CD127^{-}$ which was linked to the probability of recurrence of the cancer in 2 patients. Another clinical study reported significantly higher cytokines (TNFα and IFN-y) levels, increased T cell response to autologous tumour tissue (IFN-y ELISPOT assay) after 4 weeks and higher cytotoxic activity of T cells after 4 and 8 weeks (measured by luciferase assay) after cryoablation in 20 high-risk prostate patients. In patients with metastatic hormone refractory prostate cancer, a combination of cryoablation and GM-CSF showed a 70% decrease of PSA levels and a median time to progression of 18 months. No correlation was seen between the increase tumour-specific T cell responses in the peripheral blood and the increased cytolytic activity (measured by luciferase assay) after 4 and 8 weeks. The addition of cryoablation to androgen deprivation therapy (ADT) in 30 prostate cancer patients with bone metastases significantly improved progression-free, cancer-specific and overall survival compared to 30 prostate patients only treated with ADT.

No clinical trials have been performed so far to evaluate a combination of anti-CTLA-4 antibodies with cryoablation therapy in humans. Only immunotherapy with GM-CSF has been investigated. Currently, a phase II trial of the combination of pembrolizumab and cryosurgery in stage IV prostate patients is ongoing. In addition, other trials are ongoing searching for the relation between cryoablation and the effect on the immune system.

Lung Cancer

Percutaneous local ablative therapies are considered viable options for the treatment of stage IA non-small-cell lung carcinoma (NSCLC). Recurrent lesions after radiation therapy or surgery and metastatic lesions can be treated by means of ablation as well.

Cryoablation of lung lesions is associated with lower pain levels and fewer complications in tumours located close to the chest wall and mediastinum or central lesions close to the hilum; however, no clinical randomised studies have been executed comparing the different percutaneous ablative therapies.

NSCLC is a heterogeneous group of cancers which is known for high numbers of tumour-specific mutations that are linked with response to immunotherapy. In recent years, several immunotherapies have been approved for the treatment of lung cancer, namely PD-1 checkpoint inhibitors nivolumab and pembrolizumab and the anti-CTLA-4 inhibitor ipilimumab.

Preclinical work in a mouse model revealed that Intratumoural injection of DCs with cryoablation elicits a Th1 response with higher levels of IFN-y and effector memory CD8+ T cells observed from spleen cells resulting in protection against secondary tumours and prolonged survival. In another study, the addition of CpG ODN to the cryoablation plus DCs resulted in a significant reduction of new tumour growth, fewer metastasis development and a prolonged survival compared to all the therapies alone. A decrease in Tregs and increase in cytotoxic T cells were observed and linked to the better response in the combination group. In another study, the same treatment combination showed a higher elevation of CD4+ and CD8+ T cells and IL-12, IFN-y and TNFα together with a delay in tumour growth and improved survival in mice treated with the combination therapy. Takahashi et al. reported the greatest immune response (higher numbers of specific T cells and higher levels of stimulating cytokines) and the slowest tumour growth after two cycles of cryoablation compared to one or three cycles of cryoablation.

Clinical work combining cryoablation and allogenic intravenous NK cells showed an improvement in the quality of life and tumour response rates compared to cryoablation alone. Additional phase II/III trials must be conducted to reveal the potential benefits in larger patient groups before combination treatment is considered an alternative. Another strategy, consisting of a combination of cryoablation with gefitinib, an inhibitor of epidermal growth factor receptor's (EGFR) tyrosine kinase domain, showed significant improvement of overall response with a higher 1-year survival rate in patients treated with gefitinib and cryoablation compared to gefitinib alone. Lastly, 166 metastatic NSCLC patients received either cryoablation alone, cryoablation followed by immunotherapy (DC-CIK) or chemotherapy or all three therapies. The survival of patients treated with cryoablation combined with chemo or immunotherapy was longer than treatment consisting of chemo or immunotherapy alone (18 and 17 months vs 8.5 and 12 months).

The overall survival in patients that received the triple combination therapy (cryoablation, immunotherapy and chemotherapy) was significantly longer (27 months) compared to other groups.

These studies show positive results for the combination of cryoablation and stimulants to the immune system, NK cell therapy or DC-CIK, with improvement in survival. Clinical results are expected from a phase II study where cryoablation is combined with an immune checkpoint inhibitor.

Melanoma

Cryoablation is used for the treatment of benign superficial lesions, such as actinic keratosis, but it is not indicated as a treatment for primary melanoma. Only in unresectable lesions with high metastatic load, cryoablation may diminish tumour load by ablation of the primary site. The treatment of melanoma metastasis can be performed with cryoablation to slow down the rate of tumour spread. In metastatic mouse models, combinations with different immunostimulants (including TLR 9 and CPG) have shown an enhanced effect of cryoablation for suppression of new tumour growth. A pilot study observed the induction of endogenous heat-shock protein after administration of GM-CSF and radiofrequency ablation or cryoablation in metastatic melanoma patients, with a small number of subjects demonstrating the combination therapy as a feasible and safe therapeutic option. The combination of cryoablation and immunotherapy may be beneficial in the metastatic stetting to overcome the limitations of the immunotherapy. Currently, two trials are open combining immunotherapy and cryoablation.

CRYOIMMUNOTHERAPY

Cryoimmunotherapy, also referred to as cryoimmunology, is an oncological treatment for various cancers that combines cryoablation of tumor with immunotherapy treatment. In-vivo cryoablation of a tumor, alone, can induce an immunostimulatory, systemic anti-tumor response, resulting in a cancer vaccine—the abscopal effect. Thus, cryoablation of tumors is a way of achieving autologous, in-vivo tumor lysate vaccine and treat metastatic disease. However, cryoablation alone may produce an insufficient immune response, depending on various factors, such as high freeze rate. Combining cryotherapy with immunotherapy enhances the immunostimulating response and has synergistic effects for cancer treatment.

Although, cryoblation and immunotherapy has been used successfully in oncological clinical practice for over 100 years, and can treat metastatic disease with curative intent, it has been ignored in modern practice. Only recently has cryoimmunotherapy been resurrected to become the gold standard in cancer treatment of all stages of disease.

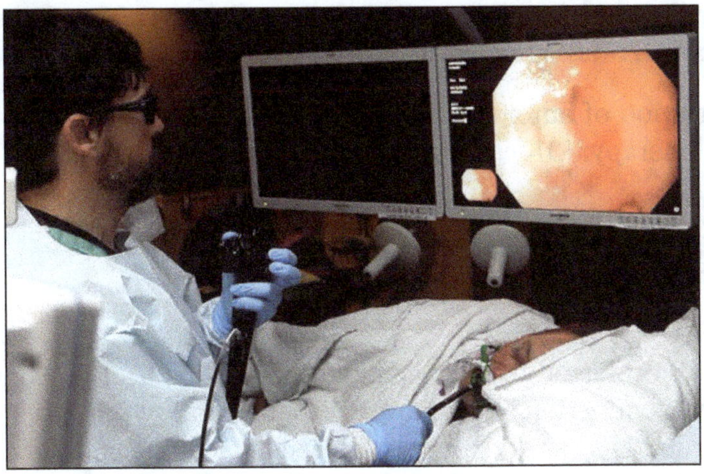

Immunological effects resulting from the cryoablation of tumors was first observed in the 1960s. Since the 1960s, Tanaka treated metastatic breast cancer patients with cryotherapy and reported cryoimmunological reaction resulting from cryotherapy. In the 1970s, systemic immunological response from local cryoablation of prostate cancer was also clinically observed. In the 1980s, Tanaka, of Japan, continued to advance the clinical practice of cryoimmunology with combination treatments including: cryochemotherapy and cryoimmunotherapy. In 1997, Russian scientists confirmed the efficacy of cryoimmunotherapy in inhibiting metastases in advanced cancer. In 2000s, China, following closely with the exciting developments, enthusiastically embraced cryoablation treatment for cancer and has been leading the practice ever since with cryoimmunotherapy treatments available for cancer patients in numerous hospitals and medical clinics throughout China. In the 2010s, American researchers and medical professionals, started to explore cryoimmunotherapy for systemic treatment of cancer.

Mechanisms of Actions

Cryoablation of tumor induces necrosis of tumor cells. The immunotherapeutic effect of cryoablation of tumor is the result of the release of intracellular tumor antigens from within the necrotized tumor cells. The released tumor antigens help activate anti-tumor T cells, which destroy remaining malignant cells. Thus, cryoablation of tumor elicits a systemic anti-tumor immunologic response.

The resulting immunostimulation from cryoablation may not be sufficient to induce sustained, systemic regression of metastases, and can be synergised with the combination of immunotherapy treatment and vaccine adjuvants.

Various adjuvant immunotherapy and chemotherapy treatments can be combined with cryoablation to sustain systemic anti-tumor response with regression of metastases, including:

- Injection of immunomodulating drugs (i.e.: therapeutic antibodies) and vaccine adjuvants (saponins) directly into the cryoablated, necrotized tumor lysate, immediately after cryoablation.

- Administration of autologous immune enhancement therapy, including: dendritic cell therapy, CIK cell therapy.

CRYONEUROLYSIS

Cryoneurolysis, also referred to as cryoanalgesia, is a medical procedure that temporarily blocks nerve conduction along peripheral nerve pathways. The procedure, which inserts a small probe to freeze the target nerve, can facilitate complete regeneration of the structure and function of the affected nerve. Cryoneurolysis has been used to treat a variety of painful conditions.

Medical Uses

A similar procedure that uses radiofrequency energy for back pain appears to have short term benefit, but it is unclear if it has a long term effect.

Mechanisms of Action

Nerve Anatomy

Each nerve is composed of a bundle of axons. Each axon is surrounded by the endoneurium connective tissue layer. These axons are bundled into fascicles surrounded by the perineurium connective tissue layer. Multiple fascicles are then surrounded by the epineurium, which is the outermost connective tissue layer of the nerve. The axons of myelinated nerves have a myelin sheath made up of Schwann cells that coat the axon.

Nerve Injury Classification

Classification of nerve damage was well-defined by Sir Herbert Seddon and Sunderland in a system that remains in use.

Cryoneurolysis treatments that use nitrous oxide (boiling point of -88.5 °C) as the coolant fall in the range of an axonotmesis injury, or 2nd degree injury, according to the Sunderland classification system. Treatments of the nerve in this temperature range are reversible. Nerves treated in this temperature range experience a disruption of the axon, with Wallerian degeneration occurring distal to the site of injury. The axon and myelin sheath are affected, but all of the connective tissues (endoneurium, perineurium, and epineurium) remain intact. Following Wallerian degeneration, the axon regenerates along the original nerve path at a rate of approximately 1–2 mm per day.

Cryoneurolysis differs from cryoablation in that cryoablation treatments utilize liquid nitrogen (boiling point of -195.8 °C) as the coolant, and therefore, fall into the range of a neurotmesis injury, or 3rd degree injury according to the Sunderland classification. Treatments of the nerve in this temperature range are irreversible. Nerves treated in this temperature range experience a disruption of both the axon and the endoneurium connective tissue layer.

Reversible	
1st Degree Neuropraxia – Interruption of conduction; Short recovery time	+10 to -20°C
2nd Degree Axonotmesis – Loss of continuity of the axon; Wallerian degeneration; Preservation of endo- peri- and epineurium	-20°C to -100°C
Non Reversible	
3rd/4th Degree Neurotmesis – Loss of continuity; Some loss of continuity of epineurium and perineurium; endoneurium may or may not be disrupted	-140°C and colder
5th Degree Transection (Severe Neurotmesis) – Gross loss of continuity	*Not possible with cryoneurolysis*

Nerve injury classification table of temperatures required for injury.

Devices

Cryoprobe

Cryoneurolysis is performed with a cryoprobe, which is composed of a hollow cannula that contains a smaller inner lumen. The pressurized coolant (nitrous oxide, carbon dioxide or liquid nitrogen) travels down the lumen and expands at the end of the lumen into the tip of the hollow cannula. No coolant exits the cryoprobe. The expansion of the pressurized liquid causes the surrounding area to cool (known as the Joule-Thomson effect) and the phase change of the liquid to gas also causes the surrounding area to cool. This causes a visible iceball to form and the tissue surrounding the end of the cryoprobe to freeze. The gas form of the coolant then travels up the length of the cryoprobe and is safely expelled. The tissue surrounding the end of the cryoprobe can reach as low as -88.5 °C with nitrous oxide as the coolant, and as low as -195.8 °C with liquid nitrogen. Temperatures below -100 °C are damaging to nerves.

Other Devices

The Endocare PerCryo Percutaneous Cryoablation device utilizes argon as a coolant and can be used with 4 different single cryoprobe configurations with a diameter of either 1.7 mm (~16 gauge) or 2.4 mm (~13 gauge) in diameter.

The Myoscience Iovera° is a handheld device that uses nitrous oxide as a coolant and can be used with a three-probe configuration with a probe diameter of 0.4 mm (~27 gauge).

CRYOTHERAPY

Cryotherapy, sometimes known as cold therapy, is the local or general use of low temperatures in medical therapy. Cryotherapy may be used to treat a variety of tissue lesions. The most prominent use of the term refers to the surgical treatment, specifically known as cryosurgery or cryoablation. Cryosurgery is the application of extremely low temperatures to destroy abnormal or diseased tissue and is used most commonly to treat skin conditions.

Cryotherapy is used in an effort to relieve muscle pain, sprains and swelling after soft tissue damage or surgery. It can be a range of treatments from the application of ice packs or immersion in ice baths (generally known as cold therapy), to the use of cold chambers.

While cryotherapy is widely used, there is little evidence as to its efficacy that has been replicated or shown in large controlled studies. Its long term side effects have also not been studied.

Ice Pack Therapy

Ice pack therapy is a treatment of cold temperatures to an injured area of the body. Though the therapy is extensively used, and it is agreed that it alleviates symptoms, testing has produced conflicting results about its efficacy.

An ice pack is placed over an injured area and is intended to absorb heat of a closed traumatic or edematous injury by using conduction to transfer thermal energy. The physiologic effects of cold application include immediate vasoconstriction with reflexive vasodilation, decreased local metabolism and enzymatic activity, and decreased oxygen demand. Cold decreases muscle spindle fiber activity and slows nerve conduction velocity, therefore it is often used to decrease spasticity and muscle guarding. It is commonly used to alleviate the pain of minor injuries, as well as decrease muscle soreness. The use of ice packs in treatment decreases the blood flow most rapidly at the beginning of the cooling period, this occurs as a result of vasoconstriction, the initial reflex sympathetic activity.

Ice is not commonly used prior to rehabilitation or performance because of its known adverse effects to performance such as decreased myotatic reflex and force production, as well as a decrease in balance immediately following ice pack therapy for 20 minutes. However, if ice pack therapy is applied for less than 10 minutes, performance can occur without detrimental effects. If the ice pack is removed at this time, athletes are sent back to training or competition directly with no decrease in performance.

Cold Spray Anesthetics

In addition to their use in cryosurgery, several types of cold aerosol sprays are used for short-term pain relief. Ordinary spray cans containing tetrafluoroethane, dimethyl ether, or similar substances, are used to numb the skin prior to or possibly in place of local anesthetic injections, and prior to other needles, small incisions, sutures, and so on. Other products containing chloroethane are used to ease sports injuries, similar to ice pack therapy.

Whole Body Cryotherapy

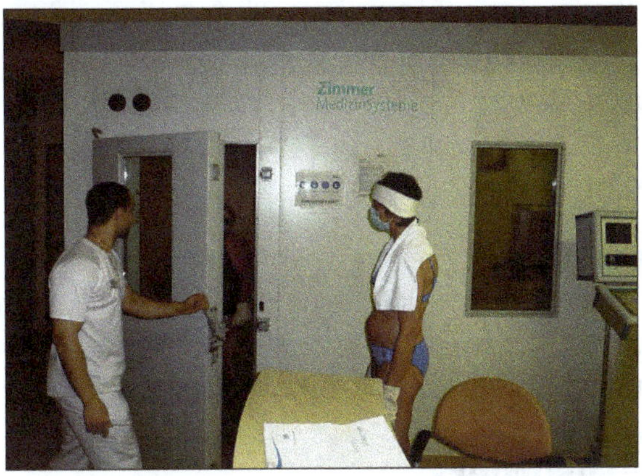

Cryotherapy patients during preparation of treatment of c. 3 minutes.

It is unclear if whole body cryotherapy (WBC) has any effect on muscle soreness, or improves recovery, after exercise. There is no evidence that whole body cooling effectively treats Alzheimer's, fibromyalgia, migraines, rheumatoid arthritis, multiple sclerosis, stress, anxiety, or chronic pain as its proponents claim.

This treatment involves exposing individuals to extremely cold dry air (below −100 °C) for two to four minutes. To achieve the subzero temperatures required for WBC, two methods are typically used: liquid nitrogen and refrigerated cold air. During these exposures, individuals wear minimal clothing, which usually consists of shorts for males, and shorts and a crop top for females. Gloves, a woollen headband covering the ears, and a nose and mouth mask, in addition to dry shoes and socks, are commonly worn to reduce the risk of cold-related injury. The first WBC chamber was built in Japan in the late 1970s, introduced to Europe in the 1980s, and has been used in the US and Australia in the past decade.

Adverse Effects

Reviews of whole body cryotherapy have called for research studies to implement active surveillance of adverse events, which are suspected of being underreported. If the cold temperatures are produced by evaporating liquid nitrogen, there is the risk of inert gas asphyxiation as well as frostbite.

Partial Body

Partial body cryotherapy (PBC) devices also exist. If the cold temperatures are produced by evaporating liquid nitrogen, there is the risk of inert gas asphyxiation as well as frostbite.

Benefits of Cryotherapy

Reduces Migraine Symptoms

Cryotherapy can help treat migraines by cooling and numbing nerves in the neck area. One study found that applying a neck wrap containing two frozen ice packs to the carotid arteries in the neck significantly reduced migraine pain in those tested. It's thought that this works by cooling the blood passing through intracranial vessels. The carotid arteries are close to the skin's surface and accessible.

Numbs Nerve Irritation

Many athletes have been using cryotherapy to treat injuries for years, and one of the reasons why is that it can numb pain. The cold can actually numb an irritated nerve. Doctors will treat the affected area with a small probe inserted into the nearby tissue. This can help treat pinched nerves or neuromas, chronic pain, or even acute injuries.

Helps Treat Mood Disorders

The ultra-cold temperatures in whole-body cryotherapy can cause physiological hormonal responses. This includes the release of adrenaline, noradrenaline, and endorphins. This can have a positive effect on those experiencing mood disorders like anxiety and depression. One study found that whole-body cryotherapy was actually effective in short-term treatment for both.

Reduces Arthritic Pain

Localized cryotherapy treatment isn't the only thing that's effective at treating serious conditions; one study found that whole-body cryotherapy significantly reduced pain in people with arthritis. They found that the treatment was well-tolerated. It also allowed for more aggressive physiotherapy and occupational therapy as a result. This ultimately made rehabilitation programs more effective.

May Help Treat Low-risk Tumors

Targeted, localized cryotherapy can be used as a cancer treatment. In this context, it is called "cryosurgery." It works by freezing cancer cells and surrounding them with ice crystals. It is currently being used to treat some low-risk tumors for certain types of cancer, including prostate cancer.

May Help Prevent Dementia and Alzheimer's Disease

While more research is needed to evaluate the effectiveness of this strategy, it's theorized that whole-body cryotherapy could help prevent Alzheimer's and other types of dementia. It's thought that this may be an effective treatment because the anti-oxidative and anti-inflammatory effects of cryotherapy could help combat the inflammatory and oxidative stress responses that occur with Alzheimer's.

Treats Atopic Dermatitis and other Skin Conditions

Atopic dermatitis is a chronic inflammatory skin disease with signature symptoms of dry and itchy skin. Because cryotherapy can improve antioxidant levels in the blood and can simultaneously reduce inflammation, it makes sense that both localized and whole-body cryotherapy can help treat atopic dermatitis. Another study (in mice) examined its effect for acne, targeting the sebaceous glands.

Risks and Side Effects

The most common side effects of any type of cryotherapy are numbness, tingling, redness, and irritation of the skin. These side effects are almost always temporary. You should never use cryotherapy for longer than is recommended for the method of therapy you're using. For whole body cryotherapy, this would be more than four minutes. If you're using an ice pack or ice bath at home, you should never apply ice to the area for more than 20 minutes. Wrap ice packs in a towel so you don't damage your skin. Those with diabetes or any conditions that affect their nerves should not use cryotherapy. They may be unable to fully feel its effect, which could lead to further nerve damage.

References

- Cryosurgery-fact-sheet, about-cancer-treatment-types-surgery: cancer.gov, Retrieved 02 January, 2019
- Callstrom, Matthew R.; Dupuy, Damian E.; Solomon, Stephen B.; Beres, Robert A.; Littrup, Peter J.; Davis, Kirkland W.; Paz-Fumagalli, Ricardo; Hoffman, Cheryl; Atwell, Thomas D. (2013-03-01). "Percutaneous image-guided cryoablation of painful metastases involving bone". Cancer. 119 (5): 1033–41. doi:10.1002/cncr.27793. ISSN 1097-0142. PMC 5757505. PMID 23065947
- Hohenauer Erich, Taeymans Jan, Baeyens Jean-Pierre, Clarys Peter, Clijsen Ron (2015). "The Effect of Post-Exercise Cryotherapy on Recovery Characteristics: A Systematic Review and Meta-Analysis". PLOS ONE. 10 (9): e0139028. Bibcode:2015PLoSO..1039028H. doi:10.1371/journal.pone.0139028. PMC 4586380. PMID 26413718
- Cryotherapy-benefits-tips-and-guidelines, health: healthline.com, Retrieved 28 June, 2019
- Cooper (2001). "The History of Cryosurgery". J R Soc Med. 94 (4): 196–201. doi:10.1177/014107680109400416. PMC 1281398. PMID 11317629

Cryopreservation

CHAPTER 3

Cryopreservation uses very low temperatures to preserve various living contents such as organelles, cells, tissues, extracellular matrix, organs, or any other biological constructs. This chapter delves into cryoconservation of animal genetic resources and cryopreservation of testicular tissue, semen, oocyte, etc. to provide an easy understanding of the subject.

Cryopreservation is the preservation of cells and tissue by freezing. Cryopreservation is based on the ability of certain small molecules to enter cells and prevent dehydration and formation of intracellular ice crystals, which can cause cell death and destruction of cell organelles during the freezing process. Two common cryoprotective agents are dimethyl sulfoxide (DMSO) and glycerol. Glycerol is used primarily for cryoprotection of red blood cells, and DMSO is used for protection of most other cells and tissues. A sugar called trehalose, which occurs in organisms capable of surviving extreme dehydration, is used for freeze-drying methods of cryopreservation. Trehalose stabilizes cell membranes, and it is particularly useful for the preservation of sperm, stem cells, and blood cells.

Most systems of cellular cryopreservation use a controlled-rate freezer. This freezing system delivers liquid nitrogen into a closed chamber into which the cell suspension is placed. Careful monitoring of the rate of freezing helps to prevent rapid cellular dehydration and ice-crystal formation. In general, the cells are taken from room temperature to approximately −90 °C (−130 °F) in a controlled-rate freezer. The frozen cell suspension is then transferred into a liquid-nitrogen freezer maintained at extremely cold temperatures with nitrogen in either the vapour or the liquid phase. Cryopreservation based on freeze-drying does not require use of liquid-nitrogen freezers.

An important application of cryopreservation is in the freezing and storage of hematopoietic stem cells, which are found in the bone marrow and peripheral blood. In autologous bone-marrow rescue, hematopoietic stem cells are collected from a patient's bone marrow prior to treatment with high-dose chemotherapy. Following treatment, the patient's cryopreserved cells are thawed and infused back into the body. This procedure is necessary, since high-dose chemotherapy is extremely toxic to the bone marrow. The ability to cryopreserve hematopoietic stem cells has greatly enhanced the outcome for the treatment of certain lymphomas and solid tumour malignancies. In the case of patients with leukemia, their blood cells are cancerous and cannot be used for autologous bone-marrow rescue. As a result, these patients rely on cryopreserved blood collected from the umbilical cords of newborn infants or on cryopreserved hematopoietic stem cells obtained from donors. Since the late 1990s it has been recognized that hematopoietic stem cells and mesenchymal stem cells (derived from embryonic connective tissue) are capable of differentiating into skeletal and cardiac muscle tissues, nerve tissue, and bone. Today there is intense interest in the growth of these cells in tissue culture systems, as well as in the cryopreservation of these cells for future therapy for a wide variety of disorders, including disorders of the nervous and muscle systems and diseases of the liver and heart.

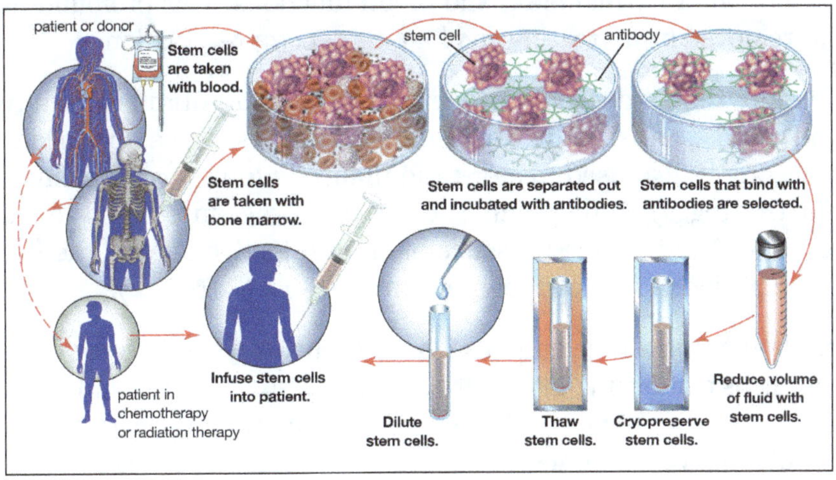

Bone marrow transplantation.

High doses of chemotherapy or radiation destroy not only cancer cells but also bone marrow, which is rich in blood-forming stem cells. In order to replace damaged marrow, stem cells are harvested from either the blood or the bone marrow of the cancer patient before therapy; cells also may be taken from a genetically compatible donor. In order to remove unwanted cells, such as tumour cells, from the sample, it is incubated with antibodies that bind only to stem cells. The fluid that contains the selected cells is reduced in volume and frozen until needed. The fluid is then thawed, diluted, and reinfused into the patient's body. Once in the bloodstream, the stem cells travel to the bone marrow, where they implant themselves and begin producing healthy cells.

Cryopreservation is also used to freeze and store human embryos and sperm. It is especially valuable for the freezing of extra embryos that are generated by in vitro fertilization (IVF). A couple can choose to use cyropreserved embryos for later pregnancies or in the event that IVF fails with fresh embryos. In the process of frozen embryo transfer, the embryos are thawed and implanted into the woman's uterus. Frozen embryo transfer is associated with a small but significant increase in the risk of childhood cancer among children born from such embryos.

Profound hypothermia, a form of mild cryopreservation used in human patients, has significant applications. A common use of induction of profound hypothermia is for complex cardiovascular surgical procedures. After the patient has been placed on complete cardiopulmonary bypass, using a heart-lung machine, the blood passes through a cooling chamber. Controlled cooling of the patient may reach extremely low temperatures of around 10–14 °C (50–57 °F). This amount of cooling effectively stops all cerebral activity and provides protection for all the vital organs. When this extreme cooling has been achieved, the heart-lung machine can be stopped, and the surgeon can correct very complex aortic and cardiac defects during circulatory arrest. During this time, no blood is circulating within the patient. After the surgery has been completed, the blood is gradually warmed in the same heat exchanger used for cooling. Gradual warming back to normal body temperatures results in resumption of normal brain and organ functions. This profound hypothermia, however, is far removed from freezing and long-term cryopreservation.

Cells can live more than a decade if properly frozen. In addition, certain tissues, such as parathyroid glands, veins, cardiac valves, and aortic tissue, can be successfully cryopreserved. Freezing is also used to store and maintain long-term viability of early human embryos, ova (eggs), and sperm. The freezing procedures used for these tissues are well established, and, in the presence of cryoprotective agents, the tissues can be stored over long periods of time at temperatures of −14 °C (6.8 °F).

Research has shown that whole animals frozen in the absence of cryoprotective agents can yield viable cells containing intact DNA upon thawing. For example, nuclei of brain cells from whole mice stored at −20 °C (−4 °F) for more than 15 years have been used to generate lines of embryonic stem cells. These cells were subsequently used to produce mouse clones.

Basic Principles of Cryopreservation

Spermatozoa were the first mammalian cells to be cryopreserved successfully. This success was due to the serendipitous discovery by Polge and co-workers of the cryoprotective effect of glycerol. Since then, many methods have been developed for various types of cells, tissues and organs. Much progress in the field has come from empirical work as well as from fundamental cryobiology. Increased understanding of the causes of cryo-injury has continually helped to improve cryopreservation methods. Research into fundamental cryobiology has provided the basis for new cryopreservation methods such as vitrification.

The two most commonly used cryopreservation methods for animal germplasm are slow-freezing and vitrification. These are quite different methods, but relate to the same physico-chemical relationships. The differences between the two can be explained by first describing what happens during slow freezing.

Slow-freezing

In slow-freezing, cells in a medium are cooled to below freezing point. At some stage, ice masses containing pure crystalline water will form. What remains between the growing ice masses is the so-called unfrozen fraction, in which all cells and all solutes are confined. The concentrations of sugars, salts and cryoprotectant (e.g. glycerol) increase, while the volume of the unfrozen fraction decreases. The increase in osmotic strength causes an efflux of water from the cells. Slow cooling is needed in order to allow sufficient efflux of water to minimize the chance of intracellular ice formation. As cooling continues, the viscosity of the unfrozen fraction ultimately becomes too high for any further crystallization. The remaining unfrozen fraction turns into an amorphous solid that contains no ice crystals.

Chilling Injury and Cold Shock

The first challenge in cryopreserving cells from homeotherm (warm-blooded) animals is in cooling the cells below body temperature. Cells may be damaged by very rapid cooling (cold shock) or be damaged by low temperature per se (chilling injury). Behaviour and function of membrane lipids and proteins may be affected by temperature. For example, membrane lipids that are normally in a liquid crystalline state may solidify at non-physiological temperatures, which can change their function and begin processes such as cryocapacitation of the production of reactive oxygen species that increase damage to membranes. Decreasing the temperature may cause an imbalance in cellular processes because the rate of one process may be affected more strongly than that of another. One example is the disintegration of the metaphase spindle of oocytes caused by a change in the dynamic equilibrium of the association/dissociation of the tubulin filaments.

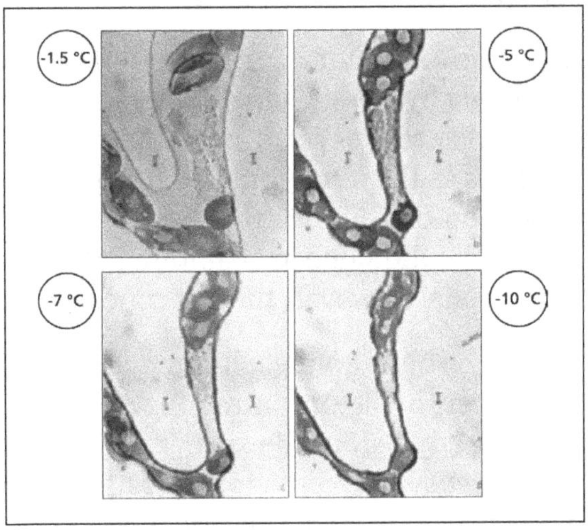

Frog erythrocytes in the "unfrozen fraction", which is enclosed by growing masses of ice.

Supercooling

In slow-freezing methods cells are brought into a suitable freezing medium and cooling is continued below the freezing point of the medium. Ice formation does not necessarily start at the freezing point. Small ice crystals have a lower melting/freezing point than "bulk" ice, due to their large surface tension. Spontaneous ice nucleation will in most cases occur after the solution is supercooled to a temperature between -5 and -15 °C. Thereafter, ice will grow rapidly in all directions, and the release of the latent heat of fusion will cause the sample to warm up abruptly until the freezing/melting temperature of the solution (i.e. of the remaining unfrozen fraction) is reached. At this point, the ice formation will stop, or will proceed at a rate governed by the rate at which the heat of fusion is transported from the sample. Finally, the sample can "catch up" again with the lower temperature in the freezing apparatus. From a practical perspective, this means that the cells undergoing cryopreservation in a typical semen straw have to withstand a series of large and abrupt temperature changes.

Conditions in the Unfrozen Fraction

Cells are faced with very high concentrations of solutes in the unfrozen fraction. Dehydration and high salt concentration may result in loss of stability in the membranes or denaturation of proteins. Moreover, high salt concentrations may cause extracellular salts to enter the cells, a process known as "solute loading".

The fast efflux of water causes a rapid decrease in the volume of the cells to approximately 50 percent of their original volume. This leads to structural deformation of the cells. Further mechanical stress may be caused by cells being confined in very narrow channels of unfrozen solution and squeezed between growing masses of ice.

The Influence of Cryoprotectants

At all practical cooling rates, the total solute concentration (which is measured in moles per kg water) is determined only by the subzero temperature. When the initial freezing medium contains only salts (electrolytes), salt concentrations in the unfrozen fraction will reach extremely high levels as the temperature decreases. In contrast, in a medium that contains a large proportion of non-electrolytes, the total solute concentration at each subzero temperature will be the same as that found at the equivalent temperature in a medium containing only salts; however, the salt concentration will be much lower.

Sugars can be used as non-electrolyte solutes, but they will only affect the extracellular salt concentration. Moreover, high concentrations of impermeable solutes impose osmotic stress on the cells already before freezing. This is much less the case when a membrane permeable solute, such as glycerol, is used rather than a non-permeable solute. When cells are brought into a hypertonic glycerol medium, water will leave the cells because of the osmotic pressure difference. However, at the same time, glycerol will enter the cells. After a short period of equilibration, the cells will have regained their original volume. The osmotic stress imposed by a hypertonic glycerol solution is therefore much smaller than that imposed by a hypertonic sugar solution. Hence, glycerol can be used at greater concentrations than sugars without damaging the cells. A substantial initial glycerol concentration in the medium means that part of the extracellular and intracellular water is replaced by the glycerol. Hence, the amount of ice formed is lower, the unfrozen fraction remains larger, the degree of shrinkage of the cells is limited, and the electrolyte concentration in the unfrozen solution and in the cells will be relatively small. The mechanisms through which other membrane permeable substances, such as ethylene glycol and dimethyl-sulfoxide (DMSO), provide cryoprotection are similar to those involving glycerol.

There are additional mechanisms through which polyols, such as like glycerol and several sugars, provide cryoprotection. These substances can stabilize lipid membranes by hydrogen bonding with the polar head groups of membrane lipids, which is especially important under severely dehydrated conditions. In addition, these substances may affect the mechanical properties of the unfrozen fraction, especially its viscosity and glass-forming tendency.

The degree to which cells shrink and re-swell after addition of a membrane-permeable cryoprotectant depends on the concentration of the cryoprotectant and the relative permeability of the membrane to water and to the cryoprotectant. For instance, bull sperm shrink very little when brought into a freezing medium with glycerol, whereas bovine embryos react much more strongly. Upon thawing, removal of the cryoprotectant has the opposite effect on cells: they first swell and then they shrink again. This may lead to damage if the cells expand too much. Damage due to over-swelling of cells can be prevented by stepwise removal of the cryoprotectant.

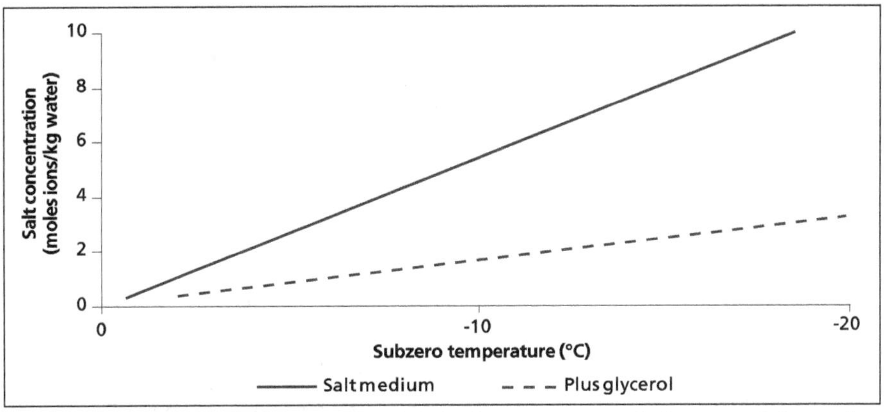

Effect of subzero on salt concentration in the presence and absence of glycerol.

The Influence of Cooling Rate

A general observation in the cryopreservation of cells and other biological systems is that each system has a specific optimal cooling rate, with decreased survival at cooling rates that are too low (slow-cooling damage) or too high (fast-cooling damage).

Ice growth is a rapid process, but transport of water through the cell membrane is relatively slow, because the membrane acts as a resistance barrier. Therefore, as cooling and extracellular ice growth continue, the liquid water of the unfrozen fraction remains very close to equilibrium with the ice, but the intracellular water lags behind. This means that the water concentration (i.e. the chemical potential of water) is too high for thermodynamic equilibrium, and there may be a risk of intracellular ice formation.

The optimal cooling rate falls in a range that is neither too fast nor too slow. When cells are cooled very slowly, the intracellular water lags behind only a little, and the risk of intracellular ice formation is minimal. However, it also means that the dehydration of the cells is maximal, which is not desired. At higher cooling rates, intracellular dehydration, intracellular solute concentration and shrinkage of the cells is less excessive. Moreover, the cells are exposed to the unfavourable conditions for a shorter period of time. However, when cooling rates are increased too much, the dehydration may not be fast enough to prevent intracellular ice nucleation. Fast-cooling damage can also be caused by other factors. For instance, it has been proposed that rapid water flow through membrane pores could lead to an uneven distribution of pressure on the membrane. Fast-cooling damage could also result from the very sudden changes in size, shape and ultrastructure, caused by the rapid efflux of water.

Different cells or other biological materials (embryos, tissue pieces) may have different optimal cooling rates. The optimal cooling rate of cells is largely determined by their volume and their membrane surface area (volume to surface area ratio), and by the permeability of the membrane to water and to cryoprotectant.

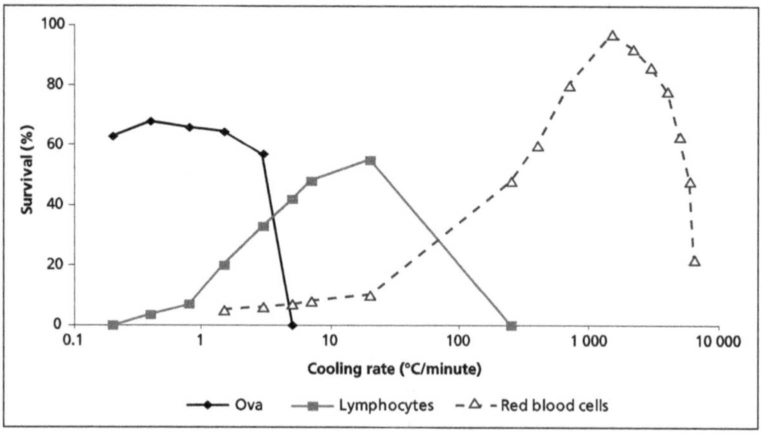

Effect of cooling rate on survival of different kinds of cells.

Interactions of Cooling Rate with Thawing Rate and Cryoprotectant Concentration

The optimal cooling rate may depend on various other factors, such as the cryoprotectant concentration and the thawing rate. It has been observed in semen from a number of species that the combination of fast cooling and slow thawing is particularly damaging to the cells. If intracellular ice nucleation occurs at a low temperature and cooling proceeds rapidly, it may be that the cytoplasm turns into glass before the intracellular ice crystals grow to a significant size, thus causing only sublethal, or no, damage. During slow thawing, the small crystals can grow and subsequently damage the cells. In addition, cells may be damaged by extracellular restructuring of ice masses, a process known as "recrystallization".

Programmable and Non-programmable Freezers

Biological material can either be frozen using quite simple, non-programmable, freezers or using more sophisticated, programmable, freezers. Although programmable freezers are more expensive, they do not necessarily yield more satisfactory results, especially for experienced technicians and cryobiologists. Therefore, the choice between programmable and non-programmable systems will depend on the financial resources available and the experience of the technicians. In some cases, even the most experienced technicians prefer the operating simplicity of programmable models.

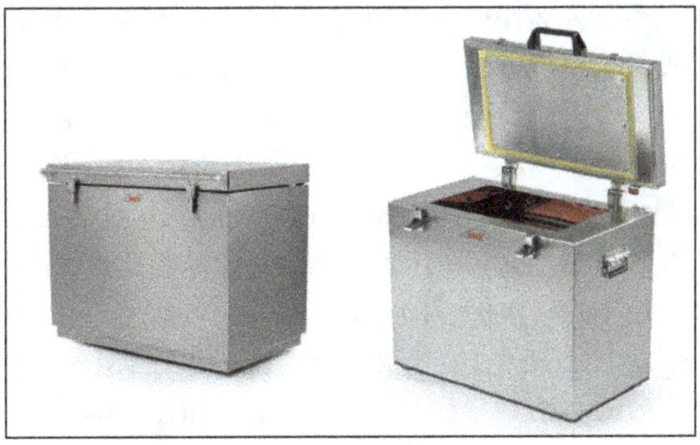

Example of a programmable freezer.

In most programmable freezers, the straws or vials are cooled by cold nitrogen vapour. The temperature inside the cooling chamber can be accurately controlled and the time course of the temperature can be programmed. However, the time course of temperature inside the straws may be different due to the generation of heat of fusion.

In non-programmable freezers, the straws may be cooled by being exposed to vapour (or a cold surface) at a constant low temperature. An example of a simple system is the freezing of straws placed on a rack in a Styrofoam box partially filled with liquid nitrogen without ventilation. The height of the straws above the liquid nitrogen determines the rate of heat exchange. Alternatively, straws can be placed on a piece of Styrofoam that floats on the liquid nitrogen. The thickness of the Styrofoam piece determines the rate of heat exchange.

Generally in such systems, the rate of heat exchange is governed by the temperature difference between the inside and the outside of the straw, and by the extent of heat conduction. The latter depends strongly on the volume to surface ratio of the straw or vial and the rate of (forced) ventilation. Therefore, it is difficult to compare one type of non-programmable freezer with another, or to know the actual freezing rate obtained with any given non-programmable apparatus. Optimal conditions have to be determined by experimentation.

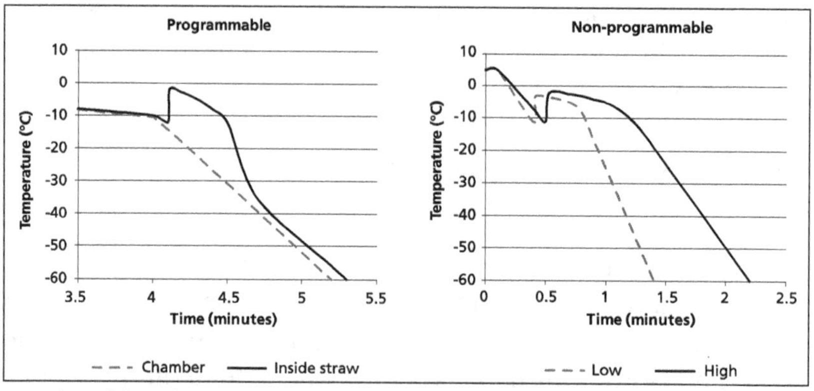

Examples of freezing curves in programmable and non-programmable freezers.

Non-programmable systems do, however, have an advantage. The cooling curve (the time course of cooling and freezing) is, by default, of the form theoretically predicted to be optimal for slow freezing, with relatively low cooling rates directly after ice formation begins and higher cooling rates later. The bulk of the ice formation happens in the temperature range between the freezing point and -10 °C, and consequently most of the water efflux from the cells must also take place in this temperature range. Thus, the heat of fusion liberated during ice formation slows the cooling exactly at the point when cells need extra time to export intracellular water. The overall steepness of the freezing curve can be adjusted in non-programmable systems by choosing the height of the straws above the liquid nitrogen, which is proportional to the temperature of the vapour around the straws. In more sophisticated systems with forced ventilation and adjustable preset vapour temperatures, the rate of heat exchange can be adjusted by choosing the preset vapour temperature.

Vitrification

The term "vitrification" refers to any process resulting in "glass formation", the transformation from a liquid to a solid in the absence of crystallization. According to this definition, cells that are properly slow frozen become "vitrified".

If, in slow-cooling methods, cells ultimately become vitrified, how do so-called vitrification methods differ? Vitrification methods involve the use of a medium that has a very high solute concentration to begin with. Thus, ice cannot form in any part of the sample. As no ice forms, cooling does not have to be slow. In fact, it may be beneficial to cool very rapidly. The vitrified state and the associated physico-chemical conditions obtained using vitrification methods, are to some extent similar to those obtained by slow cooling, but the way of reaching this point is quite different.

Chilling Injury and Cold Shock

As in the case of slow-freezing methods, vitrification methods can damage cells or tissues through cold shock and chilling injury. Depending on the material and the protocol used, however, cells or tissues may be rapidly cooled from a temperature at which chilling injury and cold shock play no role (e.g. room temperature). Extremely high rates of cooling from such a temperature to the vitrified state seem to be able to "outrun" cold shock and chilling injury. For example, rapid cooling seems to prevent disintegration of the metaphase spindle of oocytes.

Cryoprotective Agents

In vitrification methods, cells or tissues are brought into a medium that has a very high concentration of cryoprotective agents, also known as cryoprotectants. If the concentration of solutes is high enough, vitrification solutions will solidify to a glass without any risk of intracellular or extracellular ice formation during cooling or warming, independently of the cooling and warming rates used. However, the very high concentrations of cryoprotective agent needed for vitrification may cause damage due to abrupt osmotic changes, extremely low water potential or chemical toxicity. According to the description provided by Rall, the embryos are first equilibrated with 25 percent vitrification solution at room temperature. Then the embryos are cooled to 4 °C and transferred to 50 percent vitrification solution and then to 100 percent vitrification solution. They are then rapidly packed and transferred into liquid nitrogen. The stepwise increase of cryoprotective agent concentration reduces osmotic effects, while the low temperature and rapid transfer help prevent damage by chemical toxicity. In addition, chemical toxicity may be reduced by using mixtures of various permeant CPAs, or addition of non-permeant CPAs (60 g/litre polyethylene glycol) or 60 g/litre bovine serum albumin (BSA).

Reduction of Cryoprotective Agent Concentration at High Cooling Rates

Solutions that have a solute concentration lower than that of classical vitrification solutions have freezing points below which there is a significant tendency to form ice crystals. But when the solution is cooled very rapidly, there is simply no time for ice formation. Below a certain temperature, the solution becomes so viscous and stiff that ice formation becomes impossible, and the solution turns into "metastable" glass. The solute concentration needed for metastable vitrification decreases as a function of increasing cooling rate. The most recent vitrification procedures, therefore, make use of high cooling rates in order to reduce the concentration of CPAs and thereby decrease the damage caused by osmotic stress and chemical toxicity.

The cooling rate can be increased in several ways. One is to reduce the volume of the sample to be vitrified. An early example of this approach is the open pulled straw method (often abbreviated OPS). Even smaller sample volumes have been used on electron microscope grids, so-called hemi-straws, nylon loops (cryoloops) or polypropylene strips (Cryotop – Kitazato Supply Co., Fujinomiya, Japan). The Cryotop system allows a volume of 0.1 µl to be vitrified.

In addition to reducing the sample volume, a faster cooling rate can be achieved by heat transfer to a liquid that does not boil. Liquid nitrogen at its boiling point (-196 °C) will generate nitrogen gas when it absorbs heat. This will create a film of gas that insulates the sample from the liquid nitrogen. Liquid nitrogen at its freezing point (also known as "nitrogen slush") doesn't have this disadvantage. It can be produced with an apparatus called Vit Master.

In metastable vitrification procedures, it is also essential that the warming (i.e. thawing) of the sample is very rapid. If warming is slow, ice crystals can form while the temperature is between the vitrification temperature and the freezing point of the vitrification solution.

Most recent vitrification protocols make use of these ultra-rapid approaches in order to reduce cryoprotective agent concentrations and prevent cold shock and chilling injury. Current vitrification solutions have much lower solute concentrations than those used in classical vitrification solutions. As described in Section 4, very good results are currently obtained when using these approaches for vitrification of oocytes and embryos. Recent studies with pig and cattle oocytes have indicated that the Cryotop system gives better results than the open pulled straw system.

Freeze-drying

Storage of freeze-dried biological material is extremely cost efficient, as no expensive and bulky liquid nitrogen containers are necessary. Furthermore, it is safe. The material may be stored at ambient temperature and, unlike cryogenic storage, there is no risk of equipment malfunction or of personal injury from liquid nitrogen. On the negative side, however, freeze drying generally reduces cell viability. Therefore, standard insemination procedures generally cannot be used for freeze-dried sperm. However, freeze-dried sperm have been successfully used to produce live offspring using ICSI in mice and rabbits. In addition, freeze-dried somatic cells have been successfully used to produce apparently healthy embryos using SCNT. However, there have so far been no reports of cloned offspring produced by SCNT using freeze-dried somatic cells. Thus, while freeze drying is potentially useful for gene banking of genetic resources with the objective of regenerating live animals and recovering lost breeds, this would require further development and optimization of procedures. Conversely, freeze-dried gametes and somatic cells can already be used for conservation of germplasm intended for use in (genetic) research.

The key to freeze-drying is to bring the material to a vitrified glass state in which the glass transition temperature is higher than ambient temperature. The first step is to bring the biological material to a vitrified state. The next step is to apply a vacuum to the material, which results in sublimation of any ice that may be present and further decreases the water content of the vitrified material. This increases the glass transition temperature, which ultimately reaches a level higher than the ambient temperature.

Thus, at the end of the process the material can be stored at ambient temperature while remaining in the stable glass state. Obviously, the initial freezing/vitrification procedure, and the medium used, should be optimized so as to ensure the survival of the germplasm throughout this phase. In addition, the medium composition must be optimized so as to prevent the cells from being damaged by the effects of the further dehydration of the material.

PRECISION IN CRYOPRESERVATION

For any cryopreservation protocol there are five key questions that govern the methodology and logistics of the freezing and storing process.

- What is to be stored?
- How many batches are to be stored?
- What is the expected duration of storage?
- What properties are the retrieved samples required to possess?
- Are there packaging requirements in addition to those dictated by the cryopreservation process?

Reduction of temperature results in the retardation of metabolic processes and this can, in some circumstances, provide sufficient stability for the required period of storage. However, at temperatures below 0 °C the biological effects of cooling are dominated by the crystallization of ice: typically, water constitutes around 80 % of tissue mass. Freezing is the conversion of liquid water to crystalline ice but the term is commonly misused in circumstances where samples are cooled below their expected freezing point but without the formation of ice, for example by supercooling or by vitrification. The result of the freezing of water in a complex solution is that the concentration of the solutes in the remaining liquid phase increases and some solutes may precipitate if their concentration exceeds their solubility limits. This realisation provides two potential mechanisms of damage: direct mechanical effects of the formation of ice, and the rise in concentration of dissolved solutes.

In 1948 a method was discovered that permitted the freezing of many types of animal cells with good post-thaw recovery of living cells: Adding 10-20 % of glycerol enabled avian spermatozoa to survive freezing at -80 °C. Theories of freezing injury that were current at the time envisaged ice crystals damaging the cells and intracellular structures, and because glycerol increased the total solute concentration in the system, the amount of ice that formed was reduced. A little later, in the 1950s, Lovelock (1952) showed that the increase in concentration of salts as the volume of the suspending solution decreased was in fact the dominant damaging mechanism: salt concentration, rather than ice formation, was a major cause of freezing injury to cells. Subsequently other cryoprotectant solutes were explored along with different rates of cooling, resulting in solidification of the stored samples but with a range of mixtures of ice and vitrified solid in the stored samples.

The physical nature of the sample dictates the thermal transfer characteristics of the cooling process for that specific sample and either the physical size or cell-type will affect the appropriate cooling rate and other parameters of the cryopreservation protocol. Similarly, the physical type and ultimate intended use of the sample (for example dose requirement in the case of future therapeutic use) will determine the size of the individual packaging. An additional layer of packaging may be necessary to prevent microbiological contamination – so-called 'double bagging'. Likewise the ultimate destination of the sample will also dictate the care required during the freezing process and the conditions necessary for long-term storage. Some tissues and most larger biological samples are currently difficult or impossible to cryopreserve successfully and new techniques, such as Liquidus Tracking may address some of the problems associated with cryopreservation of these types of sample.

It is sometimes the case that the ultimate use of the samples stored is not known at the time of the initial collection and storage and sometimes the significance of particular samples may change with time. However, in many cases, the potential of the stored samples is fixed or limited at the time of selecting the cryopreservation and storage methods. The importance of these choices will be covered later; however it is pertinent to note here that the storage process may have an important impact on the value of samples when they are recovered from storage; changes in the properties of the recovered samples may be irreversible and this is therefore a key to maximising the sample's potential.

The term "viability" is frequently used in the context of cell and tissue banking. Strictly speaking it means the potential to exhibit the signs of life at some future stage, whereas it is often misused to mean the extent to which a sample demonstrates attributes of life at the present time. But that is "vitality" not "viability." However, it is also the case that not all the attributes of life are exhibited by all living things, and the possession of one attribute does not imply the presence of them all. In fact, few of the properties that characterise "life" can be measured quantitatively. The term is best avoided; functional measurements should be named to describe what they actually measure: membrane integrity; a specified metabolic function; ability to reproduce. In addition there are obvious cases where the tissue does not have to be alive in order to function; in bone for example. But equally, in many cases fully functional survival is paramount; the haemopoietic stem cells in cord blood will not graft in the recipient if the cell concentration is lower than a threshold value. In such cases a low total recovery of living cells in the thawed sample will limit the use of the thawed sample. Another common situation is where samples are stored in order to ensure that a supply of identical cells will be available throughout a long-term study. Although it is possible to regrow new cell batches from recovered samples, repeating this process can lead to progressive degradation due to mutations.

The Physics of Freezing

The process of freezing is ultimately simple; it is merely the application of an environment that removes energy from the sample over a period of time and changes the physical state of water in the sample from liquid to crystalline. Crystalline water (ice) excludes the solutes previously dissolved in the water, resulting in two potentially dangerous mechanisms – direct effects of ice and secondary effects in the solute composition. At a sufficiently low temperature all biological activity is prevented and the physical state of the sample is preserved. In simple cases, where the only requirement is to preserve the physical state or where cellular structure is absent (viruses, DNA etc.), that is the end of the story; physical deterioration can be prevented at relatively high temperatures, and in many institutions worldwide this task is completed in the banks of -80 °C refrigeration units that proliferate in medical and biological research establishments.

The preservation of living cells and tissues and the post-thaw ability of cells to proliferate and thrive are determined by a number of factors: the laboratory techniques and the thermodynamic processes that a sample experiences during processing and freezing; the environment in which it resides between freezing and the ultimate use post-thaw. The potential of many samples is severely limited at this stage by the choices made by, or enforced upon, the technician regarding the freezing protocol. It may be that some stored samples lose significant value due simply to the omission of a few simple additional steps.

The cryopreservation process has two main aims. The first is to reduce the temperature of the sample to a point where biological stability is achieved. The application of an external cryogenic environment will remove energy from the sample and create a very low-energy solid state within which biological and chemical activity are limited or prevented altogether. The second is that during the freezing process it is necessary to prevent the formation of intracellular ice crystals: such crystals damage the cellular structure and can lead to limited post-thaw recovery and post-thaw failure of the cell sample to function as required. Additionally, the protocol must take into account the stresses to which the cells are exposed during the freezing process (dehydration, hypothermia, chemical toxicity, and solute concentration) and the potential for an apoptotic response post-thaw.

The objective, therefore, is to create an environment in which, as the sample is cooled, the chemical composition inside the cell, is managed in such a way as to create an intracellular composition with a lower freezing point than the applied environment, whilst maintaining an external suspending composition that is able to solidify at the same temperature. The balance between the internal and external environment is managed chemically via the solutes in the micro-environment and thermodynamically via the application of an energy reducing (cooling) macro-environment. It is the combined action of these two factors that determines the success or otherwise of a cryopreservation protocol for the conservation of vitality.

The appropriate solute composition is created by including cryoprotective agents (CPAs) in the medium. These operate in one of two ways: Either they modify the extracellular composition or alternatively they also replace some of the intracellular water. The first mechanism involves the addition of non-penetrating CPAs such as trehalose, polyethylene glycol (PEG) or polyvinyl-pyrrolidone (PVP), to the medium. The second mechanism requires the addition of penetrating solutes that can traverse the cell membrane, such as glycerol, ethylene glycol and dimethyl sulphoxide (DMSO). Since water does not retain solutes when it freezes, a solution at equilibrium with ice will vary in osmotic potential as it freezes and because of this, the micro-environment of a cell will require either the cell to lose water to the environment or exchange water for CPA molecules, thereby maintaining osmotic balance. The concentration of intracellular material lowers the effective freezing point of intracellular material and, provided the external temperature is correctly managed, prevents the formation of intracellular ice. As such, the creation of ice crystals within the cell is avoided. At temperatures below -130 °C (close to the glass transition temperature of the medium) the residual liquid has too little energy to orientate into long range molecular matrices and will form short range semi-solid structures; i.e. an amorphous solid or glass. At this point there is no possibility for significant chemical transport; biological activity, and hence deterioration, effectively ceases.

The options for control of this process are the chosen CPA and its concentration, and the cooling rate. Water and solute permeability are temperature dependent and nominally the higher the concentration of extracellular CPA, the less ice will form during cooling. With a very high applied concentration of CPA, very rapid cooling without the formation of ice may be possible – a process that is known as vitrification. At the other extreme, lower CPA concentrations that allow ice to form, require more precisely managed cooling rates which can be provided by programmable controlled rate freezers. The issue here is the toxicity of the applied CPA since high concentrations, even for short periods, can lead to excessive dehydration and high cell stress, whereas lower concentrations may involve prolonged cellular exposure to essentially toxic material. DMSO, for example, is an organic solvent and has been linked to cellular mutation. The choices made for the preparation and subsequent freezing of cells is a complex balance between thermodynamic and biochemical variables, the choice and management of which can have a profound effect on the post-thaw recovery of living cells and hence the value of the sample.

Long Term Cell Survival and Contamination

All biological materials will, without intervention, naturally deteriorate, and if they are to be preserved it is necessary to utilize a method that will preserve both morphology and functionality while preventing any alteration of the fundamental nature of the material. The most common methodology available for this is cryopreservation. Biological materials, however, have widely different properties and in order to create a truly effective cryopreservation protocol, it is necessary to consider these properties as they affect the preservation of vital characteristics both during the freezing process and the subsequent environment in which the samples are to be stored long term.

Regarding the minimum storage temperature, no temperature is too cold. Once a sample is frozen and the residual liquid phase has vitrified, further cooling simply reduces molecular energy and vibration. It is possible for short-range structural changes to occur at a molecular level, but they do not affect post-thaw biological properties. It is worth noting that because the cell micro-environment within a frozen sample is chemically different from the majority of the frozen material, biological activity may continue, albeit slowly, at temperatures several degrees below the freezing point of the material.

The minimum melting point of the multi-dimensional phase diagram for typical cryopreservation media occurs at around -80 °C but the cell contents do not finally solidify to an amorphous state until around -120 °C. It is not sufficient simply to keep the samples frozen because, at a micro-environmental level, if the material retains the ability to diffuse it may also degrade, albeit at a much reduced rate. The glass transition temperature is therefore regarded as the "critical" temperature if truly long term storage is required.

Best practice dictates that freezers should maintain sample temperatures as far as possible below this critical temperature. By storing well below the critical temperature, transitory warming events above that temperature can be avoided during sample handling, retrieval, storage and in the event of any disruption to the availability of cryogen or power. Freezing a sample in such a way as to maintain maximum biological potential is not a trivial task, and the same care applied to this process should be brought to bear when designing and building storage environments.

Clearly, the process should be able to maximise the potential for use after processing and storage. Because the future use may be unclear, the preservation and storage procedures should be designed to provide the best possible opportunity for future exploitation. The storage of cells without either adequate care during the initial cryopreservation process or at too high a temperature during subsequent long-term storage are key problems that should be avoided and when the purpose of storage is to maintain biological potential, it is vital that the mechanism of freezing injury be considered.

As the liquid in which the cells are suspended begins to freeze, any solutes in the unfrozen solution become more concentrated and this results in a depression of the freezing point of the remaining solution. The result of this, when the temperature is reduced, is that the cells are exposed to a solution of progressively higher concentration. The increasing concentration increases the osmotic gradient across the cell membrane which results in water leaving the cell in order to maintain balance. Hence, controlling the cooling rate provides a mechanism for controlled dehydration of the cells. Eventually the aqueous phase is so viscous that there is insufficient energy available for the water molecules to form a crystalline solid and the solution becomes an amorphous solid or glass. The temperature at which this condition is reached is known as the "glass transition" temperature (Tg). Once the sample is below this temperature, diffusion within and without the cell stops and the sample is biologically inert. At temperatures below Tg the sample can be maintained indefinitely; other physical interactions, such as background radiation, may have an impact on extremely long-term storage but such effects are probably without any significance in practice.

Unlike freezing, the glass transition is not based upon a thermodynamically defined phase change but rather on the observed dramatic change in viscosity that occurs in cryoprotectant solutions typically at around -120 °C. It is important to ensure that samples are maintained below this temperature throughout the storage term. A temperature of -150 °C is typically stated to be the critical storage temperature for cell products since this temperature provides a reasonable safety margin to ensure that that samples remain below the glass transition temperature during transitory events such as handling, but in practice, storage in liquid nitrogen at -196 °C is a convenient and reliable way to meet this requirement; moreover the additional safety margin provides even greater sample security.

However, storage in liquid nitrogen is not without its disadvantages which include the risk of explosion during warming should liquid nitrogen have entered the vials. Microbiological cross-contamination is another hazard of storage in liquid nitrogen and may lead to the application of a secondary enclosure ('double bagging'). Storage in the gas phase has been advocated to avoid these problems. In the past, the temperature gradient in the vapour phase of liquid nitrogen refrigerators has been a problem, and there may have been increased vulnerability to inadequate amounts of liquid nitrogen between refills. Modern high-efficiency liquid nitrogen cooled vessels now allow storage in the vapour phase without these problems. These vessels are vacuum insulated and the surface area that is not insulated by the vacuum is minimised ensuring that the evaporation rate of liquid nitrogen is kept low. Restricting the amount of energy entering the vessel ensures that the temperature in the vapour phase is maintained close to the liquid nitrogen temperature. The upper region of the refrigerator, close to its access point where the temperature would otherwise be higher, can be efficiently cooled if it is ensured that the heat exchange surface extends right from the bottom to the top of the refrigerator: gas phase temperatures of around -190 °C can be achieved.

The weak point in the process of maintaining safe low temperatures for samples focuses on the time in transport to and from its storage. Small samples of low thermal mass, such as vitrified straws, can warm at the rate of thousands of degrees Celsius per minute and regulatory inspections requiring the removal of samples for identification can be another weak point.

Traceability

Under most regulatory environments, a rigorous sample tracking system is a key and mandatory component of compliance. It is vital that the individual location of any sample is recorded accurately, and that the sample is labelled with a unique identifier such that the identity of a sample at any location can be verified. For many research and therapy provision operations it is also necessary to have all processing, analytical and, if relevant, patient data linked in a central database.

Labelling can be a challenge as sample containers can be small and the cryogenic environment hostile; however commercially available cryogenic-proof labels and label printing systems are readily available. RFID tags are also a promising solution.

Sample location databases should be organized hierarchically, such that the location of any individual sample can be readily identified; for example: Room/Freezer/Shelf; or Segment/Rack/Position of Box in Rack/position of Vial in Box; or Room/Freezer/Canister/Cane/Goblet/Straw position in Goblet.

Most regulatory environments require the label to include both machine and human readable identifiers (bar code plus text) and where a sample is stored in a secondary container (such as a blood bag in a cassette) it is vital that both the external container and the primary sample container be correctly labelled.

Concomitant with good identification procedures are good location and retrieval methods and there are a number of commercial software systems available with varying degrees of sophistication to accommodate larger or smaller numbers of stored samples. However an often overlooked part of the storage process is the logging, monitoring and associated alarms. Recording the parameters of storage is sometimes seen only as a regulatory obligation but liquid nitrogen levels or temperatures and the performance of mechanical freezers is of front-line importance. Alarms that work in practice rather than in theory are vital additions to a comprehensive storage environment.

Types of Technique

There are various options to consider when choosing the methodology and equipment for the cooling process. In conventional cryopreservation, where the intention is to control the rate of formation of ice in the material, it is necessary either to vary the rate of application of a cryogen when working against a constant warm environment, or to provide insulation or energy while maintaining a constant external cold environment. If insulation is used, the cooling rate at any point is approximately proportional to the difference in temperature between the sample and the environment as modified by the insulation and the change in specific heat of the sample as cooling proceeds. Thus, during the process, the cooling rate asymptotically approaches zero as thermal equilibrium is achieved. Applying variable energy to a sample in a cold environment allows the rate of cooling to be modified during the process. The aim is to maintain a composition within the cells that varies as cooling proceeds such that its freezing point remains below the applied environmental temperature. Alternatively, if the concentration of cryoprotectant is high enough, it may be possible to cool the sample sufficiently rapidly that ice cannot form – an approach called vitrification. The required cooling rate will depend on the cryoprotectant and its concentration, the latter being dependant on the concentration that the cells will tolerate. In general, very rapid but uncontrolled cooling is used. The new technique of liquidus tracking allows slow cooling and vitrification.

Freezing in Mechanical Freezers

Passive cooling uses insulation to moderate the cooling rate of samples that have been equilibrated with low concentrations of cryoprotectant and then placed inside traditional electromechanical refrigerators at -20 °C, -40 °C, -80 °C or even at lower temperatures. The cells are dessicated slowly during the cooling process. This method can be used for most robust cells but even under the best circumstances the post-thaw recovery rates may not be ideal. In addition, in most cases no instrumentation monitoring or recording of the process is provided. The variation of temperature within mechanical refrigerators is well known with one study reporting values of -43.5 °C to -90 °C in upright freezers. Since there is no active control during the process, it follows that the poor repeatability of the process can affect the cooling rate and hence the efficiency of the whole procedure. Variability might be improved if the local environment were more stable and protected from instantaneous variation due to external factors such as door openings etc. It is generally preferable to use a liquid nitrogen gas phase freezer for this approach since the internal temperature variation is small and the environment disturbed less frequently.

This approach to cooling and freezing is increasingly being used for material provision in pharmaceutical drug screening programmes as they move from supply by continuous culture towards a "cells-as-reagents" concept. In this approach, the cells are insulated in polystyrene containers as they are cooled initially to -80 °C and then transferred for cryogenic storage into liquid or vapour phase nitrogen. The need for rapid use of the cells for drug assays following cryopreservation, places an increased importance on the postthaw quality of the cells. In particular, temperature fluctuations within the polystyrene container and the storage time at -80 °C can significantly impact the post-thaw recovery of the cells and their biological function.

Controlled Freezing, Protocols and Seeding

Liquid nitrogen may be applied via a pressurised supply and cryogenic valve to create a very accurate cooling profile of temperature over time. This methodology offers the most options for optimization since the cooling rate can be varied at multiple stages in the process. As freezing proceeds the concentration of solutes in the medium increases causing cell dehydration in the sample.

Cooling protocols are designed to manage the intracellular solute concentration. The key point is the nucleation temperature of the suspending medium - that is, the temperature at which ice starts to form. The ice is extracellular, resulting in an increase in the extracellular solute concentration and and hence an osmotic pressure difference between the intracellular and the extracellular solutions that leads to the withdrawal of water from the cells. It is important to recognize that under normal circumstances, solutions do not freeze at their freezing point; they freeze at their nucleation temperature, which is variable and depends on the availability of nucleation centres in the sample. The nucleation temperature is normally several degrees below the nominal freezing point.

Once the extracellular fluid begins to freeze, two major events occur. First, as explained above, the concentration of CPA increases in the fraction of the extracellular fluid that has not at this point frozen, and this causes the cells to dehydrate. Secondly, the temperature of the suspension where freezing has commenced rises towards the nominal freezing temperature and remains at or close to this temperature until the freezing process is complete. This is followed by a drop in temperature as the sample catches up with the temperature of the surrounding medium, but if the cooling rate is too rapid the intracellular CPA concentration may be insufficient to prevent intracellular freezing – with severe consequences for the cells.

In order to avoid this hazard, the control program may be designed to allow equilibration of the sample and its suspending medium at a temperature marginally below the calculated freezing point and at this temperature the sample forced to begin to freeze by applying either a physical nucleation point via a cold instrument placed on the external wall of the sample container, or via a sudden, short-lived introduction of cryogen into the environment. This causes the sample to commence freezing. As the sample was originally held only marginally below the nominal fusion temperature, the cell experiences a much more moderate reduction in temperature when the fusion is complete and the temperatures re-equilibrate. After this, the cooling processes is started and continues with a temperature program that is designed to effect the necessary concentration changes to maintain the intracellular composition in the liquid region of the phase diagram. This process is called "seeding".

Warming and Thawing

It is usual to thaw cryopreserved or vitrified samples rapidly – typically by plunging them in a 37 °C water bath. The warming rate does have an effect on the recovery of living cells but this is not as great an influence as cooling rate is during cooling. In fact, optimum cooling rates have usually been determined using rapid warming so it is hardly surprising that rapid warming then gives the highest recovery. However, there are circumstances when the warming rate is of importance in its own right. The first is when the sample has been vitrified but is nucleated without a significant amount of ice being present. This is an unstable situation and in such circumstances the warming must be rapid to avoid intracellular freezing during warming. This consideration argues for rapid warming. The other situation occurs when the frozen material contains a significant amount of vitrified material, as is always the case in conventional cryopreservation. Glasses are brittle and the hazard here is that rapid warming will generate thermal stresses and cause the vitreous material to fracture. This will not matter greatly with cell suspensions where a fracture running through the sample is unlikely to traverse many cells but it is very important when the extracellular matrix must be intact – as it must, for example, in grafted blood vessels and heart valves. The solution here is to warm through the vitreous zone, that is from -196 °C to -123 °C, relatively slowly: Once above the T_g there is no hazard from fractures and the sample can be warmed as rapidly as you like. A convenient way to do this is to allow the sample to warm slowly in a -80 °C refrigerator or packed in solid CO_2 until its temperature is at above -100 °C. Alternatively the sample can be surrounded by a layer of insulation during the initial stage of warming in room air. A warming rate of around 50 °C/minute up to -100 °C was 'slow' enough to prevent fractures in cryopreserved rabbit carotid arteries.

Liquidus Tracking – A New Method

The controlled-rate freezing process achieves its results by preventing the formation of intracellular ice. In some samples, however, even extracellular ice can be severely damaging.

An example of this is articular cartilage. Isolated chondrocytes can be cryopreserved using conventional techniques but results when attempting to cryopreserve chondrocytes in situ have proven to be very disappointing. It was found that traditional cryopreservation results in the formation of ice crystals within the chondrons and not just in the acellular matrix which might have been expected from experience with conventional cryopreservation. In articular cartilage it is important to prevent both intracellular and extracellular ice. With this requirement in mind, the most appropriate cryopreservation approach would appear to be vitrification; that is the prevention of any ice formation at all. However, it will be clear that conventional vitrification is out of the question because of the heat transfer problems with bulky samples. Liquidus tracking (LT) provides a new approach to this problem.

During conventional cryopreservation, with a moderate concentration of CPA (say 10 %w/w) and relatively slow cooling (say 1 °C/minute), the cells are exposed to gradually increasing concentrations of cryoprotectant as progressively more extracellular ice is formed. The instantaneous CPA concentration is determined by the temperature according to the phase diagram of that specific system. The idea of LT is to control the instantaneous concentration of CPA throughout the cooling process so that the CPA concentration follows the liquidus line by external control rather than by progressive freezing of the medium. In this way the medium remains just above its freezing point at all times and no ice is formed. It is important to note that the cells are exposed only to the concentrations of CPA that they would experience during conventional cryopreservation. And we know that isolated chondrocytes in suspension can be cryopreserved by standard methods. In effect, the LT process takes advantage of the decrease in cytotoxicity of cryoprotectants as the temperature is decreased: hence, rather than starting with a very high concentration of cryoprotectant, the LT approach controls the concentration dynamically throughout the cooling process. In this way, vitrification can be achieved without using the extremely high concentrations of cryprotectant at the start of the process and without the need for rapid cooling. Of course, allowance has to be made for the time that diffusion of CPA into the tissue takes and this can be very considerable. On the other hand, if an organ can be perfused with the cryoprotectant solution, via the vascular system during cooling, then the diffusion distances will be very short and mass transport delays much less significant. In practice, when designing an LT process for a particular tissue, it is crucial to determine the concentration that is actually achieved in the tissue as the process continues and to adjust the concentration/ temperature/ time program to achieve the desired tissue concentration at all stages of the process. This necessitates slow cooling, commonly of the order of 0.1 to 0.3 °C per minute. The cooling of the samples can be achieved in a conventional controlled rate cooler and the solution composition can be controlled by standard peristaltic pumps, the whole system being under computer control.

Types of Equipment

Due to the many different types of samples that can be cryopreserved and their differing sensitivities, a number of different techniques and types of equipment are used.

Minus 80 °C and Lower: Mechanical Freezers

Mechanical refrigeration always applies the same methods no matter the degree of cooling desired: a gas is passed through a compression system and liquefied. The energy which is released during this liquefaction is dissipated to the environment via heat exchanger coils. The liquid is then passed through cooling coils within the freezer chamber and absorbs energy from the chamber as it vaporizes. It is the vaporization process that creates the cooling effect. As lower temperatures are required, lower liquid point gases must be employed. In order to liquefy these gases, higher pressures are required and often the liquefaction cannot be completed in a single process; this results in larger, multiple compressors being employed.

The most commonly used freezers for cryogenic purposes are upright, front-opening freezers with a cold point at a nominal -80 °C. It should be noted that there is no biological significance for this temperature, merely a physical significance since it approximates to the sublimation temperature of dry ice (solid carbon dioxide, -79 °C). This type of freezer can be employed to store biomaterial in which living cells are not a prime concern, or when it is to be stored for only a short period of time. This type of equipment is intended to be for transactional storage – holding material required daily and which will either be consumed or transferred to more appropriate conditions within a short time - 6 to 12 weeks typically.

The front opening design, while adding considerable convenience, creates a significant issue with temperature stability and variability. Because cold air is significantly heavier than warm air, opening the door causes massive air exchanges and temperature rises in the sample area in a short period of time. In addition, because the compressor systems run on a very high cycle time, there is little spare capacity to effect a cooling after the temperature has risen and it can take some time to return to equilibrium after a warming event. This property is similarly exhibited when the freezer is in normal operation and as has been previously noted, there can be significant temperature variations. The use of deep drawers within the refrigerator for the storage of samples is helpful in reducing the loss of cold air when the door is opened.

Because of the high cycle times, compressor failures are quite common and expensive to repair. It should also be noted that as the energy removed from the sample area is 100 % dissipated into the room in which the freezer is located, the term cost of operating a unit such as this should take into account not only the electricity consumption required for the compressor system, but also the significant air conditioning costs associated with the expelled heat from the freezers. If this energy is not removed by air-conditioning, the freezers become less efficient as room temperature rises, compressors are required to cycle even longer, power usage rises and compressors fail more quickly. Environmental management at a macro as well as micro level is therefore important.

Alcohol Bath Freezers

These commonly used laboratory units are essentially refrigerated circulators. A reservoir of cooling medium (normally an alcohol) is passed through a cooling system and re-enters a reservoir, reducing the temperature. The degree of refrigeration applied and the flow rate through the cooling coils determine the derived temperature of the reservoir. The relatively large volume of cooling liquid creates two noticeable effects: temperatures are very stable due to the large heat capacity of the available fluid and cooling rates can be controlled very accurately for a similar reason. The corollary to this however is that the rates achievable are very low and so rapid (> 1 °C/minute) rates are very hard to achieve. In addition, alcohol bath freezers are normally limited to temperatures above -80 °C.

Liquid Nitrogen Vessels: Liquid and Vapour

Storage of important biomaterial in liquid nitrogen at -196 °C is widely practised. This method allows for a 70 °C plus safety zone when considering the -120 °C threshold for longterm storage. Liquid nitrogen storage does provide the greatest safety zone. However, it also presents a number of problems, including personal safety and potential microbiological cross-contamination via the liquid nitrogen.

Storage in the vapour stage is felt to address these issues but it does come with its own set of problems. The vapour is not as cold as the liquid nitrogen itself and as such the 70°C safety margin is diminished. However, modern vapour storage vessels use carefully designed vacuum insulation to minimise the heat leakage from the environment into the vessel. This allows the vessel to maintain a vapour temperature at around -190 °C resulting in samples still being maintained at a safe distance from the glass transition temperature. Efficient designs also result in very low liquid nitrogen usage and temperatures can be maintained for up to a month without additional filling; temperatures are even maintained with the lid removed for short periods.

Controlled Rate Liquid Nitrogen Freezers

As described previously, up to present times, the controlled rate freezer offers the widest control options for a freezing protocol. With a truly variable application system for cryogen, most sample sizes can be easily accommodated and rates from the very slow (< 0.1 °C/minute) to in excess of 50 °C/minute are both achievable and controllable. Sample size, container dimension, cell volume, membrane permeability etc. are all variable factors. As the controlledrate freezer allows complex, fully controlled temperature versus time profiles to be created, protocols can be designed that are appropriate to the cell type and cryoprotectant concentration. Additional steps such as pauses for manual 'seeding' or rapid plunges to initiated freezing can be added to the profile. Transition to different rates can be triggered from the chamber temperature or representative sample; triggering the transition from the sample temperature can help remove variability introduced by different sample loads.

From an instrumentation standpoint, the programming and record-keeping intrinsic within the system meet most external compliance standards and optional software packages are generally available to enhance this aspect beyond the current requirements of any legislative authority. The fact that it is possible to optimise processes for every unique cell type together with the compliance aspect lend great versatility to this type of instrument in most application areas.

Equipment for Conventional (High Cooling Rate) Vitrification

Such protocols call for extremely rapid solidification of the sample, typically by plunging it directly, and in a somewhat uncontrolled manner, into liquid nitrogen. Intracellular ice formation is avoided by the application of very high concentrations of CPA. Equipment such as the VitMaster (IMT ltd.) can be used to increase the cooling rate. This uses negative pressure to depress the freezing point of liquid nitrogen to below -205 °C thereby increasing the cooling rate. Several open techniques have been developed to minimise the sample volume and achieve high cooling rates; for example the Cryotop method which uses a thin film strip to hold the sample. These open systems typically expose the sample directly to the liquid nitrogen which assists in achieving the very high cooling rates. Of course exposing the sample directly to liquid nitrogen in this manner raises questions of potential contamination from the cryogen. Other approaches, such as the Cryologic Vitrification Method, still use an open device at the stage of vitrification but cool the sample by touching on a liquid nitrogen cooled aluminium block. This means that the sample is not directly exposed to the liquid nitrogen and the block avoids the Leidenfrost effect. Alternative approaches use closed straws. These avoid the contamination issues but at the expense of the cooling rate. By definition, the vitrification stage of the process is difficult to measure, monitor or document, so validation and on-going quality control are qualitative exercises only.

Stirling Engines

Originally conceived in 1816 by the Reverend Stirling, the Stirling engine converts heat energy into mechanical work. The principal also works the other way round to convert mechanical energy to heat, when the Stirling engine forms a heat pump able to move heat 'uphill' from a cold place to a warmer one. This gives the Stirling engine an application as a refrigeration unit.

Most refrigerators operate on the Rankine cycle which depends on refrigerants existing with appropriate boiling points. Triple stage Rankine machines are at the limit of the technology and achieve roughly -140 °C. Although the Stirling cycle is less efficient than Rankine cycle machines, it is capable of cooling to lower temperatures and therefore comes into its own below -140 °C; miniature cryo-coolers based on Stirling engines are now quite common. Due to relative inefficiency, these Stirling based cryo-coolers can normally freeze only quite small samples of a few tens of grams maximum and cannot compete with liquid nitrogen powered machines for cooling capacity. On the other hand, they excel in clean rooms where it is not possible to obtain a supply of liquid nitrogen and it is only desired to freeze very small samples.

Liquidus Tracking Equipment

Because liquidus tracking is a relatively novel technique, there is little choice of equipment to assist with research into its use. Planer plc do manufacture a Liquidus Tracking controller that can be used for research into this approach. The equipment comprises a conventional slow-rate chamber coupled with a liquidus tracking controller and two peristaltic pumps.

The controller cools the sample in a similar manner to the conventional slow-rate freezing process. The cooling profile is typically a simple linear ramp. During the cooling of the sample, the controller monitors the current chamber temperature and adjusts the speeds of the two pumps to dynamically alter the concentration of cryoprotectant surrounding the samples. In the ideal process the concentration of cryoprotectant is maintained just above the liquidus curve. As the temperature decreases, the concentration of cryoprotectant is therefore increased. However, as the temperature of the sample decreases, the toxicity of the cryoprotectant decreases and this allows the sample to tolerate the ever increasing concentrations.

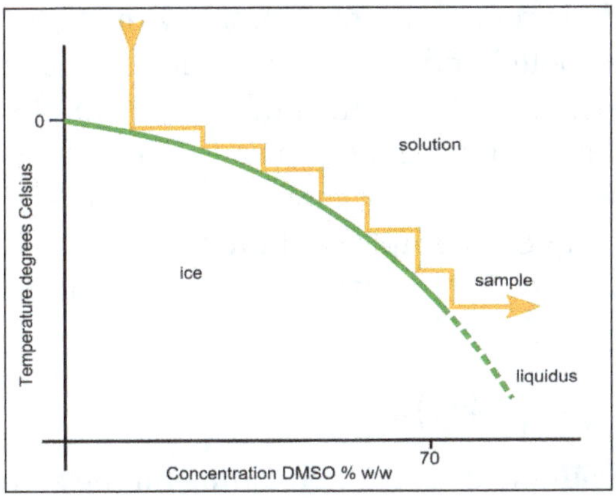

Concentration of sample tracking liquidus.

Two specific requirements of this process are the rather large volumes of cryoprotectant required and the need to ensure good mixing around the sample. The Planer Liquidus Tracker supports two modes of operation each with its own advantages and disadvantages; these are the single solution and dual solution modes.

In the dual solution mode, a solution containing a high concentration of CPA (typically 72 % w/w DMSO plus isotonic salts) and a solution containing only isotonic salts (nominal 0 % solution) are used. Each solution is pumped through a mixing junction and a heat exchanger into the sample container and thence into a waste collection container. The relative speeds of the pumps are continuously adjusted via a computer program to deliver the correct concentration to the sample. The pump speeds are adjusted to maintain a constant flow rate through the sample container. The dual solution system requires a small volume surrounding the sample so that the incoming, premixed solution is able to displace the existing solution completely as it flows through the container. The total volume of cryoprotectant can be quite large; for example, a run from 0 °C to -70 °C at 0.3 °C/minute requires a total volume of solution equal to 233 times the sample container volume. This method is suitable for use with small sample containers and has been used for discs of ovine articular cartilage.

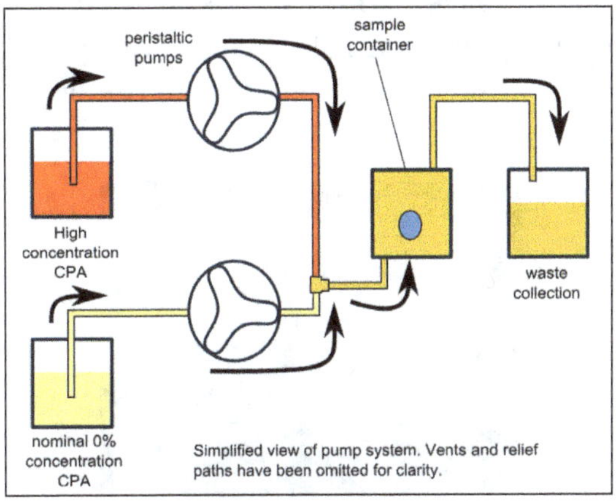

Dual solution mode.

The single solution mode is more suitable for use with larger samples. Here a highly concentrated solution is cooled and delivered to the sample container. This increases the concentration of the CPA solution surrounding the sample. To maintain a constant volume within the container, the second pump extracts the excess solution from the container. This technique is suitable for larger samples as it reduces the total volume of cryoprotectant required. For a sample container volume of Vs, a cryoprotectant concentration of Ksol and a target concentration of Kt, the volume of cryoprotectant V can be calculated from this equation:

$$V = Vs.\ln\left(Ksol/\left(Ksol - kt\right)\right).$$

For a sample container volume of 50 ml, depending on the actual values of Ksol and Kt this approach could require less than 100 ml of concentrated DMSO solution. Because the incoming solution has to be thoroughly mixed within the container, additional stirring equipment running at cryogenic temperatures is required. This results in a mechanically more complex arrangement than the dual solution approach.

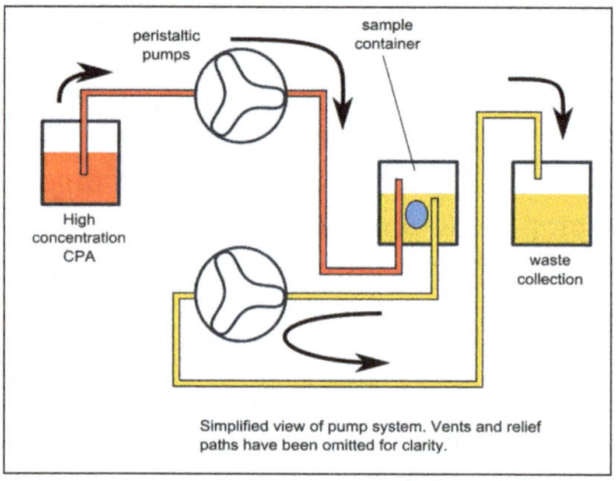

Single solution mode.

Examples of Equipment Used

The BioCool Controlled Rate Freezer from FTS Systems/SP Scientific is a mechanically refrigerated bath with temperature control to -40 °C or -80 °C. The fluid in the 2 litre bath provides temperature stability and dispersion of the heat of fusion without a concomitant temperature rise.

BioCool Controlled Rate Freezer.

The Asymptote EF600, a unique liquid nitrogen free, controlled-rate freezer, is electrically powered by a Stirling Cycle Cryocooler rather than liquid nitrogen. This allows the freezer to be used where liquid nitrogen is in short supply, where extra high air quality is needed, or where there is a risk of LN2 contamination to samples.

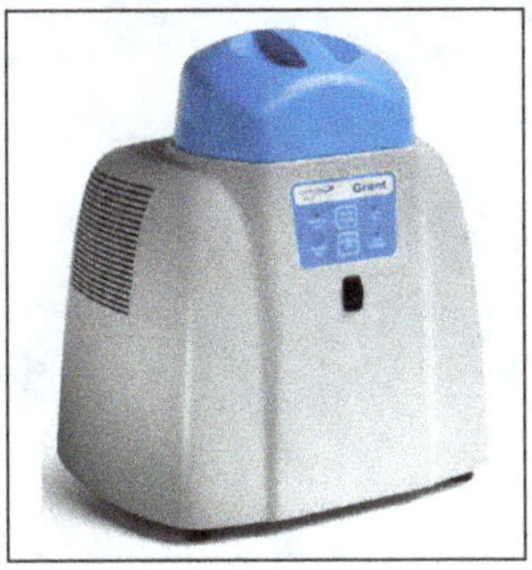

Asymptote liquid nitrogen free controlled-rate freezer.

The Planer Kryo 360 cell freezer controls down to a -180 °C end temperature to ensure sample integrity during transfer to storage. Fully programmable, it allows the use of protocols associated with the most advanced cryopreservation techniques and is widely used in laboratories around the world.

Planer Kryo 360 programmable cell freezer.

The Crysalys controlled rate freezer: programmes, time and temperatures may be entered via a touch screen with up to 100,000 cycles held on the onboard SD card; the data can be retrieved by any PC or Mac computer. A battery back up operates the system for 3 hours; its portability and 3.2 kg weight make it especially suited to veterinary purposes.

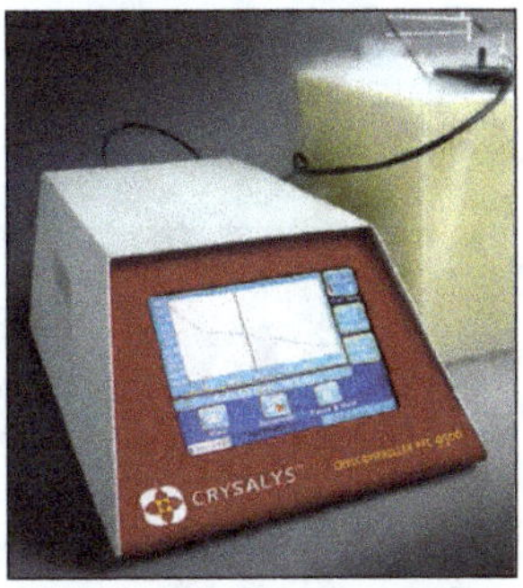

Crysalys controlled rate freezer.

The Gemini Tinytag View 2 data logger, when used with a specially designed probe, is used for temperature monitoring in cryogenic environments down to -200 °C.

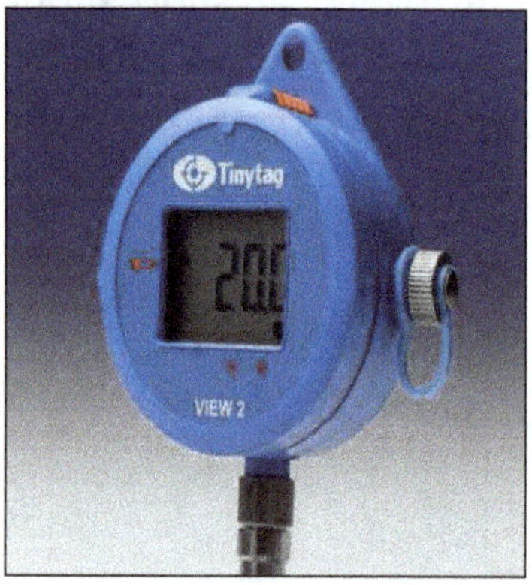

Tinytag View 2 data logger.

The CoolCell is an alcohol-free cell freezing container which provides a reproducible cooling rate of 1 °C/minute when placed in a -80 °C freezer. No alcohol is required to control the freeze rate as the design and materials of the CoolCell ensure precise and uniform heat removal from cryovials.

CoolCell, an alcohol-free cell freezing container.

The Planer ShipsLog is a datalogger specifically designed for vapour shippers, which maintains a downloadable temperature history of samples during transit.

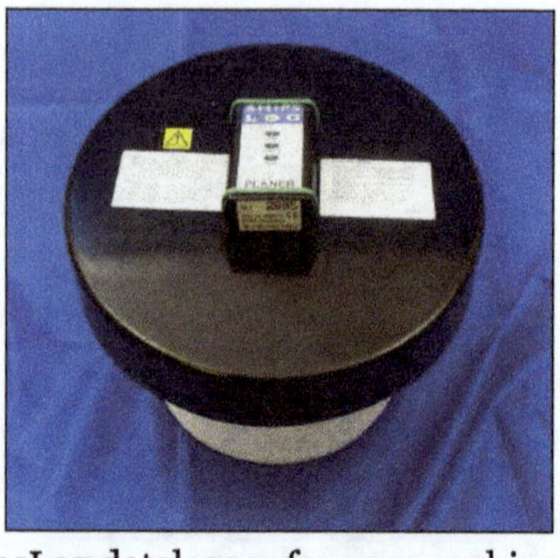

ShipsLog datalogger for vapour shippers.

The Liquidus Tracker is a new controlled vitrifier for cryopreservation of samples using the liquidus tracking technique. This approach may have uses in vitrifying larger samples and those which are currently difficult to cryopreserve.

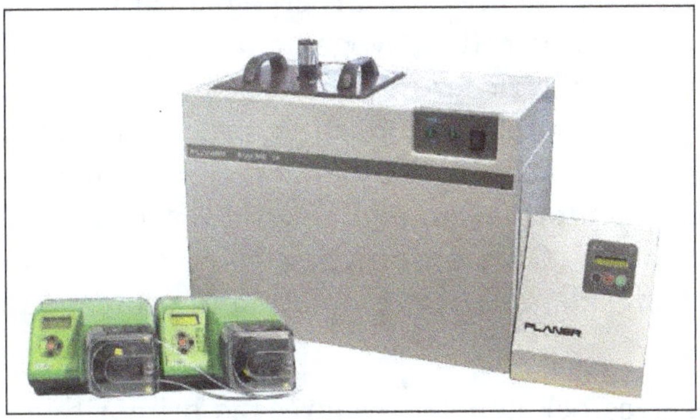

Liquidus Tracker controlled vitrifier for cryopreservation.

CRYOCONSERVATION OF ANIMAL GENETIC RESOURCES

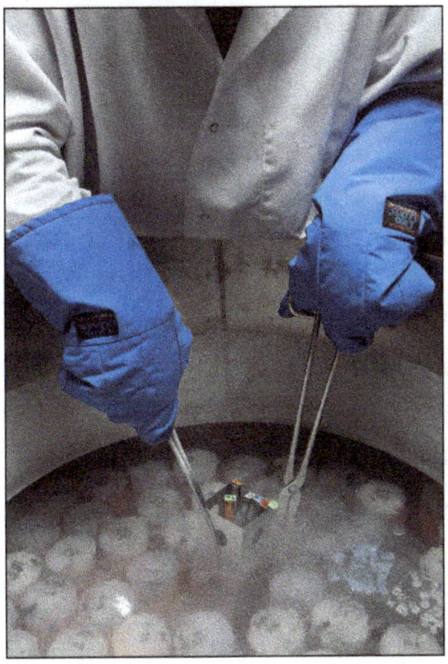

Cryoconservation of animal genetic resources at the USDA Gene Bank.

Cryoconservation of animal genetic resources is a strategy wherein samples of animal genetic materials are preserved cryogenically.

Animal genetic resources, as defined by the Food and Agriculture Organization of the United Nations, are "those animal species that are used, or may be used, for the production of food and agriculture, and the populations within each of them. These populations within each species can be classified as wild and feral populations, landraces and primary populations, standardised breeds, selected lines, varieties, strains and any conserved genetic material; all of which are currently categorized as Breeds." Genetic materials that are typically cryogenically preserved include sperm, oocytes, embryos and somatic cells. Cryogenic facilities are called gene banks and can vary greatly in size usually according to the economic resources available. They must be able to facilitate germplasm collection, processing, freezing, and long term storage, all in a hygienic and organized manner. Gene banks must maintain a precise database and make information and genetic resources accessible to properly facilitate cryoconservation. Cryoconservation is an *ex situ* conservation strategy that often coexists alongside *in situ* conservation to protect and preserve livestock genetics.

Cryoconservation of livestock genetic resources is primarily done in order to preserve the genetics of populations of interest, such as indigenous breeds, also known as local or minor breeds. Material may be stored because individuals shared specific genes and phenotypes that may be of value or have potential value for researchers or breeders. Therefore, one of the main goals remains preserving the gene pool of local breeds that may be threatened. Indigenous livestock genetics are commonly threatened by factors such as globalization, modernization, changes in production systems, inappropriate introduction of major breeds, genetic drift, inbreeding, crossbreeding, climate change, natural disasters, disease, cultural changes, and urbanization. Indigenous livestock are critical to sustainable agricultural development and food security, due to their: adaptation to environment and endemic diseases, indispensable part in local production systems, social and cultural significance, and importance to local rural economies. The genetic resources of minor breeds have value to the local farmers, consumers of the products, private companies and investors interested in crossbreeding, breed associations, governments, those conducting research and development, and non-governmental organizations. Therefore, efforts have been made by national governments and non-governmental organizations, such as the Livestock Conservancy, to encourage conservation of livestock genetics through cryoconservation, as well as through other *ex situ* and *in situ* strategies. Cryogenic specimens of livestock genetic resources can be preserved and used for extended periods of time. This advantage makes cryoconservation beneficial particularly for threatened breeds who have low breed populations. Cryogenically preserved specimens can be used to revive breeds that are endangered or extinct, for breed improvement, crossbreeding, research and development. However, cryoconservation can be an expensive strategy and requires long term hygienic and economic commitment for germplasms to remain viable. Cryoconservation can also face unique challenges based on the species, as some species have a reduced survival rate of frozen germplasm.

Cryoconservation is the process of freezing cells and tissues using liquid nitrogen to achieve extreme low temperatures with the intent of using the preserved sample to prevent the loss of genetic diversity. Semen, embryos, oocytes, somatic cells, nuclear DNA, and other types of biomaterial such as blood and serum can be stored using cryopreservation, in order to preserve genetic materials. The primary benefit of cryoconservation is the ability to save germplasms for extended periods of time, therefore maintaining the genetic diversity of a species or breed. There are two common techniques of cryopreservation: slow freezing and vitrification. Slow freezing helps eliminate the risk of intracellular ice crystals. If ice crystals form in the cells, there can be damage or destruction of genetic material. Vitrification is the process of freezing without the formation of ice crystals.

Value

Cryoconservation is an indispensable tool in the storage of genetic material of animal origin and will continue to be useful for the conservation of livestock into the future. Cryoconservation serves as a way to preserve germplasms, which is particularly beneficial for threatened breeds. Indigenous livestock may be conserved for a variety of reasons, including the preservation of local genetics, their importance in local traditions and their value to the culture identity and heritage of the area. The loss of regional livestock diversity could increase instability, decreases future possibilities and challenge production systems. Moreover, the maintenance of indigenous breeds can aid in the preservation of traditional lifestyles and livelihoods, even providing income through cultural tourism. Indigenous breeds can contribute to local economies and production systems by utilising land that is unsuitable for crop production to produce food products, as well as providing hides, manure and draft power. Therefore, the conservation and progression of these breeds are of the utmost importance for food security and sustainability.

Another beneficial factor in cryoconservation of indigenous livestock is in terms of food security and economic development. Indigenous livestock often have beneficial traits related to adaptation to local climate and diseases that can be incorporated into major breeds through cryoconservation practices. Cryoconservation is a favorable strategy because it allows germplasms to be stored for extended periods of time in a small confined area. An additional benefit of cryoconservation is the ability to preserve the biological material of both maternal and paternal cells and maintain viability over extended periods of time. Cryoconservation has been successfully used as a conservation strategy for species and breeds that have since been endangered. One drawback is that cryoconservation can only be done if preparation has taken place in advance. With proper preparation of collecting and maintaining genetic material, this method is very beneficial for the conservation of rare and endangered livestock. Cryoconservation can serve as a contingency plan when a breed population needs to be restored or when a breed has become extinct, as well as for breed improvement. This process benefits companies and researchers by making genetic materials available.

Conservation Goals			
Flexibility of country's AGR to meet changes	Insurance against changes in production conditions	Safeguarding against diseases, disasters, etc.	Opportunities for genomic research
Genetic Factors	Allowing continued breed evolution/genetic adaption	Increasing knowledge of phenotypic characteristics of breed	Minimizing exposed to genetic drafts
Sustainable utilization of total areas	Opportunities for development in rural areas	Maintenance of agro-ecosystem diversity	Conservation of rural culture diversity

The support of numerous stakeholders makes this process possible in the establishment and operations of cryoconservation. Before every phase is executed, all participating stakeholders must be briefed to understand the possible phase impending. This would include informing the stakeholders of their responsibilities and receiving their consent for the cryoconservation process. The possible stakeholders within the cryoconservation process could include:

- The State: The government acquires responsibility for conservation of animal genetic resources.
- Individual Livestock Keepers and Breed Associations: Individual livestock keepers are commonly the primary owners of the livestock whose germplasm is used for processes of cryoconservation. Breed Associations would be interested in the well-being of their respective breeds in short and long terms. Through this interest these associations may provide financial and organizational support for the cryoconservation process.
- Private Companies: Including, but, not limited to, commercial breeding companies, processing companies and agricultural support services may find value in the cryoconservation process and may striving to become more involved.
- The National Coordinator for the Management of Animal Genetic Resources: This particular stakeholder would possibly a member of the National Advisory Committee on Animal Genetic Resources. This member needs to be knowledgeable about all aspects and activities of cryoconservation, as this stakeholder would have the responsibility of reporting current information to the FAO.

Methods

Collection

There are several ways to collect the genetic materials based on which type of germplasm.

Semen

Freezing semen is a commonly used technique in the modern animal agriculture industry, which is well researched with established methods. Semen is often collected using an artificial vagina, electroejaculation, gloved-hand technique, abdominal stroking, or epididymal sperm collection. Preferred collection techniques vary based on species and available tools. Patience and technique are keys to successful collection of semen. There are several styles and types of artificial vaginas that can be used depending on the breed and species of the male. During this process the penis enters a tube that is the approximate pressure and temperature of the female's vagina. There is a disposable bag inside the tube that collects the semen. During this process it may be beneficial to have a teaser animal—an animal used to sexually tease but not impregnate the animal—to increase the arousal of the male. Electroejaculation is a method of semen collection in the cattle industry because it yields high quality semen. However, this process requires the animal to be trained and securely held, thus it is not ideal when working with wild or feral animals. When performing this process the electroejaculator is inserted into the rectum of the male. The electroejaculator stimulates the male causing an ejaculation, after which the semen is collected. The glove hand collection technique is used mainly in the swine industry. During this process, the boar mounts a dummy, while the handler grasps the penis of the boar between the ridges of his fingers and collects the semen. Abdominal stroking is exclusively used in the poultry industry. During the technique, one technician will hold the bird, while a second technician massages the bird's cloaca. However, feces and semen both exit the male bird's body through the cloaca, so the semen quality is often low.

Embryo

Embryo collection is more demanding and requires more training than semen collection because the female reproductive organs are located inside of the body cavity. Superovulation is a technique used in order to have a female release more oocytes than normal. This can be achieved by using hormones to manipulate the female's reproductive organs. The hormones used are typically gonadotropin-like, meaning they stimulate the gonads. Follicle stimulating hormone is the preferred hormone in cattle, sheep and goats. While in pigs, equine chorionic gonadotropin is preferred. However, this is not commonly done in the swine industry because gilts and sows (female pigs) naturally ovulate more than one oocyte at one time. Superovulation can be difficult because not all females will respond the same way and success will vary by species. Once the female has released the oocytes, they are fertilized internally—in vivo—and flushed out of her body. In vivo fertilization is more successful than in vitro fertilization. In cattle, usually 10 or more embryos are removed from the flushing process. In order to flush the uterus, a technician will first seal off the female's cervix and add fluid, which allows the ovum to be flushed out of the uterine horns and into a cylinder for analysis. This process typically takes 30 minutes or less. Technicians are able to determine the sex of the embryo, which can be especially beneficial in the dairy industry because it is more desirable for the embryo to be a female. Vitrification is the preferred method of embryo freezing because it yields higher quality embryos. It is crucial technicians handle the embryos with care and freeze them within 3–4 hours in order to preserve viability of the greatest percentage of embryos.

Oocytes

Oocytes can be collected from most mammalian species. Conventional oocyte collection is when ovaries are removed from a donor animal; this is done posthumously in slaughter facilities. The ovaries are kept warm as they are brought back to a laboratory for oocyte collection. Keeping the ovaries warm helps increase the success rate of fertilization. Once collected the oocytes are assessed and categorized into small, medium, and large, and then matured for 20–23 hours. This simple, inexpensive technique can lead to about 24 oocytes collected from a bovine. Conventional oocyte collection is especially useful for females who unexpectedly die or who are incapable of being bred due to injury. A second option for oocyte collection is to utilize the transvaginal ultrasound guided oocyte collection method otherwise known as TUGA. Collection technique varies slightly by species, but the general methods for collection are the same; a needle is inserted into each ovarian follicle and pulled out via vacuum. The major benefit of using this method is the ability to expand the lifetime reproductive productivity, or the number of productive days an animal is in her estrous cycle. Pregnant cows and mares continue to develop new follicles until the middle of pregnancy. Thus, TUGA can be used to substantially increase the fitness of an individual because the female then has the potential produce more than one offspring per gestation.

Somatic Cells

Somatic cells are an additional resource which can be retrieved for gene banking, particularly in the cases of emergency wherein gametes cannot be collected or stored. Tissues can be taken from living animals or shortly after death. These tissues can be saved via cryopreservation or dehydrated. Blood cells can also be useful for DNA analysis such as comparing homozygosity It is recommended by the FAO that two vials of blood be drawn to reduce the chance that all samples will be lost from a particular animal. DNA can be extracted using commercial kits, making this an affordable and accessible strategy for collecting germplasms.

	Semen	Semen and Oocytes	Embryos
Number of samples needed to restore a breed	2000	100 of each	200
Backcrossing needed?	Yes	No	No
Mitochondrial genes included?	No	Yes	Yes
Collection Possible in livestock species	Mostly, not always	Yes, in some species. Operational for bovines	Yes, in some species. Operational for bovines
Cost of collection	$$	$$	$$$$

Cryopreservation possible?	Yes	Still in experimental stage	Operational in bovines, horses and sheep. Promising in pigs. Impossible in poultry
Utilization	Surgical or non-surgical insemination backcrossing for 4 generations	In vitro maturation/IVF followed by surgical or non-surgical ET	Surgical or non-surgical ET
Current feasibility	High	Medium	High depending on available resources

Freezing

There are two cryopreservation freezing methods: slow freezing and vitrification.

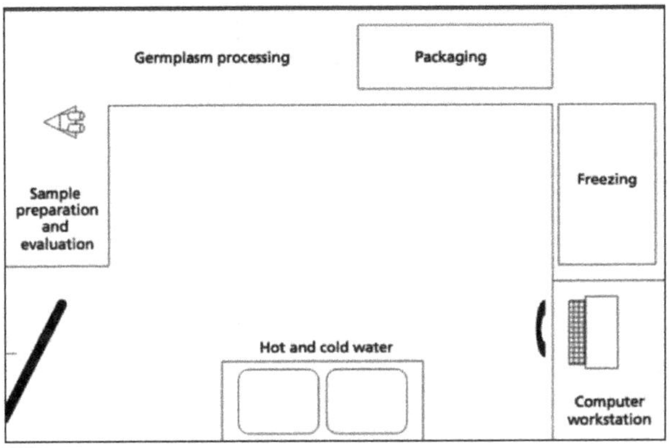

Example freezing laboratory.

Slow Freezing

During slow freezing, cells are placed in a medium which is cooled below the freezing point using liquid nitrogen. This causes an ice mass to form in the medium. As the water in the medium freezes, the concentration of the sugars, salts, and cryoprotectant increase. Due to osmosis, the water from the cells enters the medium to keep the concentrations of sugars, salts, and cryoprotectant equal. The water that leaves the cells is eventually frozen, causing more water to diffuse out of the cell. Eventually, the unfrozen portion—cellular—becomes too viscous for ice crystals to form inside of the cell.

Vitrification

The second technique for cryoconservation is vitrification or flash freezing. Vitrification is the transformation from a liquid to solid state without the formation of crystals. The process and mechanics of vitrification are similar to slow freezing, the difference lying in the concentration of the medium. The vitrification method applies a selected medium which has a higher concentration of solute so the water will leave the cells via osmosis. The medium is concentrated enough so all of the intracellular water will leave without the medium needing to be reconcentrated. The higher concentration of the medium in vitrification allows the germplasms to be frozen more rapidly than with slow freezing. Vitrification is considered to be the more effective technique of freezing germplasms.

Facility Design and Equipment

Facility Design

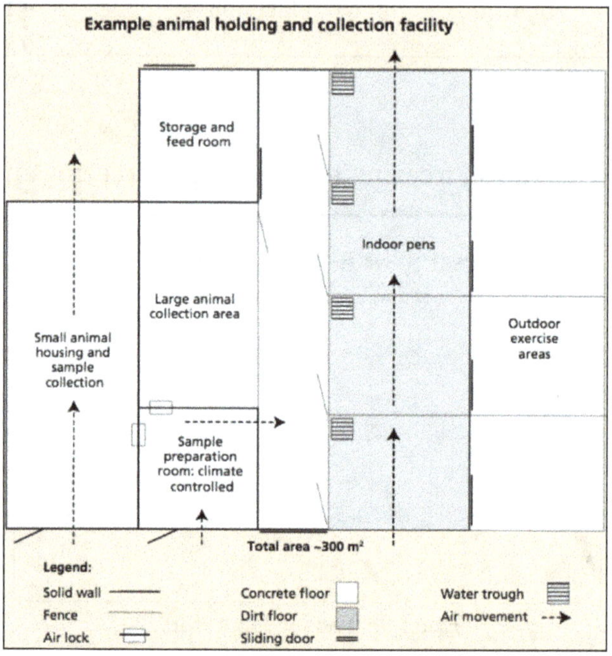

Example of animal holding and collecting facility.

When designing a facility, there are several things that should be kept in mind including biosecurity, worker safety and efficiency, and animal welfare. Diverse infrastructure is required in order to successfully collect and store genetic material. The buildings needed depend on the size of facilities as well as the extent of the operations.

Biosecurity

Biosecurity, a management measure used to prevent the transmission of diseases and disease agents on the facility, is important to keep in mind when designing a facility. In order to achieve a high level of biosecurity, collection facilities should be placed as far as possible from one another, as well as from farms. According to the FAO's recommendations, facilities should be "at least 3 km from farms or other biological risks and 1 km from main roads and railways". Separation between collection facilities and surrounding farms can improve biosecurity as pests, such as flies and mice, have the potential to travel from farm to facility and vice versa. Other disease agents may be able to travel through the air via wind, furthering the importance of separation of farms and proper air sanitation and ventilation. Additionally, a perimeter fence is used to prevent potential threats that could cause contamination to germplasms, such as unauthorized personnel or unwanted animals, from entering the facilities. Animals may be housed in pens located inside or outside of a barn as long as they are contained within the perimeter fence. When interaction with outside objects, such as feed trucks or veterinary personnel, is necessary, complete sanitation is required to decrease the risk of contamination. There is always the possibility of disease spreading among the animals whose biological data is being collected or from animal to human. An example of a disease that can easily spread through germplasm is Porcine Reproductive and Respiratory Syndrome, otherwise known as PRRS. A highly contagious disease between swine, PRRS causes millions of dollars to be lost annually by producers. The disease can be spread through boar semen. Therefore, biosecurity is particularly important when genetic material will be inserted into another animal to prevent the spread of such diseases.

Human Considerations

Worker safety is always a priority when handling livestock. Escape routes and alternative access throughout the facility are crucial for both the handlers and livestock. Germplasm storage and collection sites must include locker rooms for staff, which provide lockers, showers, and storage of clothing and footwear, in order to meet sanitation requirements.

Animal Considerations

Animal housing is practical when collecting germplasms because they keep donor animals in an easily accessible area, making the process of collecting germplasms easier and more efficient. The species and breeds of animals housed should be considered while planning the facility; facilities should be big enough to meet animal welfare standards, yet small enough to reduce human contact and increase ease of handling while reducing stress of the animal. As the process of collecting germplasm may take several days, the animal may become stressed causing a lower quality of genetic material to be obtained. Thus, training the animal to become familiar with the process is key. Holding facilities for animals may also serve as a quarantine. Quarantine facilities are necessary in order to prevent the transmission of disease from animal to animal, animal to germplasm, germplasm to germplasm, and germplasm to animal. Introducing quarantine to separate the diseased animal(s) from the healthy should be done immediately. However, a quarantine does not always prevent the spread of disease.

Temperature Control and Ventilation

Temperature control and ventilation should be included in the design of the holding and collection facilities to keep the animals comfortable and healthy, while limiting stress during the germplasm collection process. Ventilation serves as an effective way to keep clean airflow throughout the facilities and eliminate odors Temperature control helps regulate the air quality and humidity level inside the barn.

Equipment

A freezing and processing laboratory for genetic materials can be on the same site as the holding and collecting facility. However, the laboratory must have higher sanitation standards. According to the FAO, a proper germplasm laboratory should include the following.

- Washable work surfaces, floors (non-slip) and walls,
- Sufficient lighting and ventilation,
- Hot and cold, purified water,
- Electrical sockets,
- Adequate storage for consumable materials.

Cryopreservation requires equipment to collect biological material and test tubes for storage. Price is highly variable based on the quality of the collection and storage materials. The life expectancy of tools should be considered when determining costs. In addition to traditional laboratory equipment, the FAO also suggests the following:

- Disposable coveralls,
- Portable incubator,
- Haemocytometer,
- Semen straws and filling/sealing equipment,
- Liquid nitrogen storage tank,
- Liquid nitrogen,
- Liquid nitrogen dry-shipper.

Equipment for determining sperm concentration (one or more of the following three):

- Spectrophotometer (fixed or portable),
- Makler counter chamber (or disposable counting chamber),
- Haemocytometer,
- Straw filling and sealing equipment,
- Freezing equipment (manual or programmable),
- Carbon dioxide incubator (for embryos),

- Laminar flow benches (for embryos),
- Dry liquid nitrogen shipping tanks,
- Long-term liquid nitrogen storage tanks.

Limitations

Cryoconservation is limited by the cells and tissues that can be frozen and successfully thawed. Cells and tissues that can be successfully frozen are limited by their surface area. To keep cells and tissues viable, they must be frozen quickly to prevent ice crystal formation. Thus, a large surface area is beneficial. Another limitation is the species be-ing preserved. There have been difficulties using particular methods of cryoconserva-tion with certain species. For example, artificial insemination is more difficult in sheep than cattle, goats, pigs, or horses due to posterior folds in the cervix of ovines. Cryo-preservation of embryos is dependent on the species and the stage of development of the embryo. Pig embryos are the most difficult to freeze, thaw, and utilize produce live offspring due to their sensitivity to chilling and high lipid content.

Hungarian Grey Cattle

An example of the use of cryoconservation to prevent the extinction of a livestock breed is the case of the Hungarian Grey cattle, or Magya Szurke. Hungarian Grey cattle were once a dominant breed in southeastern Europe with a population of 4.9 million head in 1884. They were mainly used for draft power and meat. However, the population had decreased to 280,000 head by the end of World War II and eventually reached the low population of 187 females and 6 males from 1965 to 1970. The breed's decreased use was due primarily to the mechanization of agriculture and the adoption of major breeds, which yield higher milk production. The Hungarian government launched a project to preserve the breed, as it possesses valuable traits, such as stamina, calving ease, disease resistance, and easy adaptation to a variety of climates. The government program included various conservation strategies, including the cryopreservation of semen and embryos. The Hungarian government's conservation effort brought the population up to 10,310 in 2012, which shows significant improvement using cryoconservation.

The Gaur

Gaur, also known as the Indian bison, is the heaviest and most powerful of all wild cattle native to South and Southeast Asia. It is indicated in field data that the population of mature animals is about 5,200–18,000. Male and female Gaur both have distinctive humps between the head and shoulders, a dorsal ridge, prominent horns, and a dewlap which extends to the front legs.The Gaur grows 60% faster than domestic cattle, meaning farmers meat can be harvested at a faster rate, making beef production two to three times more profitable. Gaur meat is preferred over other breeds' meat among local people. Another benefit of the bovine is that it has the ability to sweat and tolerates heat well.

The Gaur population experienced a drastic decline of about 90% between the 1960s and 1990s due to poaching, commercial hunting, shrinking habitat, and the spreading of disease. According to the International Union for Conservation of Nature's Red List, the Gaur is a vulnerable species due to its declining population in Southeast Asia. Although the global Gaur population has declined by 30% over the past 30 years, the Gaur has a relatively stable population in India, due to protective efforts such as cryoconservation. The American Association of Zoos and Aquariums, Integrated Conservation Research (ICR), and Advanced Cell Technology have made efforts to use cryopreserved specimens of the Gaur through artificial insemination, embryo transfer, and cloning, respectively. Hybridization with domestic cattle has been successfully achieved by ICR, in order to create higher yielding, heat resistant cattle.

CRYOPRESERVATION OF TESTICULAR TISSUE

Immediate use of freshly collected testis tissue in diagnosis or in reproductive technologies is not always possible or desirable. Therefore, the ability to properly preserve the tissue for varying intervals is an essential step for maximizing the use of the source tissue. Preservation of gametes and gonads is a topic of interest in reproductive biomedicine.

Cryopreservation of human testis tissue obtained by biopsy can be used as a potential future source of sperm. For adult cancer survivors whose only source of sperm is the testis parenchyma, cryopreservation of testis biopsies may be the only option remaining if they prefer to father their own biological progeny. This will require detection of sperm in frozenthawed cell suspensions of testis tissues for use in intra-cytoplasmic sperm injection (ICSI). More importantly, cryopreservation of immature testis biopsies can offer a unique alternative for prepubertal boys undergoing gonadotoxic cancer treatments, whose only future source of spermatogenesis (i.e., spermatogonial stem cells) is at risk. These strategies can also be applied to genetic preservation of endangered species/breeds through the cryopreservation of testis tissue from young animals that die prior to reaching maturity. Restoring the developmental potential of testis tissue after cryopreservation may also provide insight into proper banking of other immature tissues.

The effects of cryoprotectant concentration and cooling rate are not similar among tissues or species.

Developmental Changes in the Structure of Testis Tissue

In mammals at birth, all organs/tissues required for sustaining life display functional competence and histological similarity to those in mature individuals. Reproductive tissues, on the other hand, attain maturity much later and only when other bodily requirements of parenthood are also in place. Therefore, in discussion of testis tissue cryopreservation, the developmental stage of the tissue is an important factor to be considered. For instance, for cryopreservation of testis tissue from an immature individual, the differing tissue texture and need for maintaining its future developmental potential are to be taken into account.

Embryonic development of the testis begins when the SRY gene in a genetic male is expressed, driving the transformation of an indifferent early gonad to a testis. This in turn causes differentiation of Sertoli cells to enclose the fetal germ cells, to mark the differentiation of primordial germ cells into gonocytes, and results in the formation of seminiferous cords. In humans, this process begins at 7-9 wk gestation and is immediately followed by differentiation of fetal Leydig cells, located in the interstitial spaces between the seminiferous cords, to allow production of testosterone thus causing masculinization of the foetus.

In early postnatal humans and most domestic species, the testis still contains interstitial tissue and seminiferous cords, with gonocytes as the only type of germ cells present. Initially, gonocytes reside in the centre of the seminiferous cords, but they gradually migrate toward the periphery of the cords and remain in close contact with Sertoli cells and peritubular myoid cells at the basement membrane to form the stem cell niche. Gonocytes eventually give rise to spermatogonial stem cells (SSCs), which have the ability to both self-renew and give rise to differentiating germ cells. Postnatal development of the testis also involves proliferation and maturation of Sertoli cells to transform testicular cords into seminiferous tubules (containing a lumen), followed by sequential division and differentiation of germ cells to generate sperm. Therefore, SSCs form the foundation of spermatogenesis and are responsible for a lifetime supply of sperm.

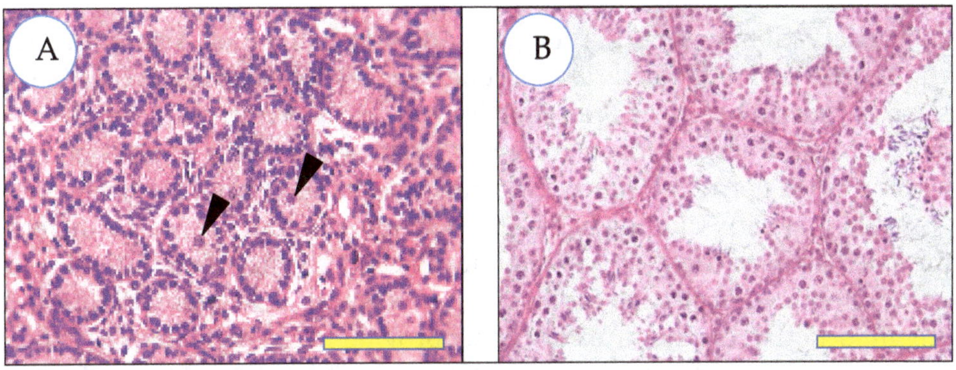

Histological differences between an immature and a mature testis tissue. In the immature testis (A), seminiferous cords contain only one type of germ cells - gonocytes (arrow heads). In the mature testis (B), on the other hand, seminiferous tubules are much larger in diameter, contain a lumen, and a repertoire of germ cell types. The composition and extent of the interstitial tissue also changes over development. These differences may affect the response of the tissue to a given cryopreservation protocol even within the same donor species. Scale bar = 100 mg.

As highlighted in the above figure, the cellular composition of a typical mature testis is quite different from that of an immature testis; for instance, the latter hosts a considerably higher number of differentiating germ cells, known to be more sensitive to manipulations and temperature changes. Consequently, the tissue composition of the testis changes during development and proportionally larger volumes of the mature testis are occupied by the seminiferous tubules. Therefore, the developmental state of the testis affects the tissue composition and has important implications for its cryopreservation.

Rationale for Preserving Testis Tissue from Human and Animal Donors

Preservation of testicular tissue could be pursued for multiple reasons. An estimated 1 in 650 children will be diagnosed with malignancies by age 16, of which 80% will be cured. However, irreversible gonadotoxic insult of chemo/radio-therapy remains a major concern in the use of these life-saving treatments, which render about 20% of boys sterile in the long term, likely as a result of the loss of spermatogonial stem cells. With improved treatments, the proportion of childhood cancer survivors is expected to increase, posing an even greater challenge for reproductive medicine and oncologist practitioners in the decades to come. A routine strategy to offer preservation of future fertility for adult men undergoing sterilizing cytotoxic treatments is to freeze semen samples; however, some men may be azoospermic at the time of cancer diagnosis. More critically, in pre-adolescent boys, collection of sperm is not possible because spermatogenesis has not yet started. In such cases, cryopreservation of testicular biopsies collected prior to the start of the treatment may provide a potential source for future use in emerging reproductive technologies.

In animal conservation, preventing the permanent loss of a male's potential contribution to the genetic variability of a rare or endangered species/breed is feasible through the collection of sperm before or even shortly after death by retrieval from the ejaculate, epididymis, or testes, which is then cryopreserved for future use in assisted reproduction. Preservation of sperm, however, is not an option when young offspring die prior to reaching sexual maturity. Cloning has been used for a number of species and especially where the goal has been to produce a genetically exact replica of an individual animal. However, development of cloning for a new species is technically demanding and costly but, more importantly, does not immediately provide the genetic diversity that would otherwise be offered by gametes. In such cases, cryopreservation of testicular tissue can again provide an alternative strategy for ex situ generation of sperm from these neonatal/immature animals for use in reproductive technologies.

Methodology for Cryopreservation of Testicular Tissue

A number of cryogenic strategies have been developed to serve as a means to maintain functional properties of the preserved cells and tissues. Apparently, the first successful cryopreservation of cells was carried out by accidental freezing of fowl sperm in diluents containing glycerol. Later, cryopreservation of bull sperm using glycerol, set the stage for revolutionizing the bovine artificial insemination industry. At about the same time, cryopreservation of unfertilized oocytes was also studied following exposure to glycerol and low temperatures. After initial success with in vitro embryo manipulation in the 1950s, research involving embryo freezing intensified. Many methods have now been developed for embryo cryopreservation and, since the 1980s, some have become routine procedures. Cryopreservation of mature oocytes has also been achieved, with high survival rates and development of normal pregnancies after in vitro fertilization (IVF).

Cryopreservation of structurally intact tissues in certain situations is more desirable than cryopreservation of isolated cells. This is especially important for complex tissues in which preservation of the target cells' functionality depends on that of other cell types present within the tissue. In case of testicular tissue, not only germ cells but also the intra-tubular supporting - Sertoli - cells as well as androgen producing interstitial - Leydig - cells are of particular interest. However, this requires devising suitable freezing protocols to maintain the existing relationship among different compartments of the tissue.

The first gonadal tissue to be successfully cryopreserved was ovarian tissue, using exposure to glycerol, resulting in preservation of cell viability and normal function after being autografted back into the animals. Subsequent reports of live rat offspring, sheep ovarian cyclic function, and pregnancy after grafting cryopreserved ovaries represented important steps in demonstrating the feasibility of this approach. Restoration of spermatogenesis was then obtained after cryopreserved testis cells were transplanted into recipient testes.

Cryopreservation of testicular tissue to be used as tissue per se, however, was not widely considered, perhaps due to lack of its potential applications. This need changed when we and others were first to show that cryopreservation of immature testis tissue prior to its xenografting can be done so as to maintain its potential for development of complete spermatogenesis. In a short period of time since then, major advances in cryopreservation of testicular tissue have opened new possibilities for preservation of male fertility in animals and humans. More recently, induction of complete spermatogenesis in vitro has further highlighted the importance of applying cryopreservation to testicular tissue for future applications. Overall, major advances have been made in the cryopreservation of reproductive tissues.

Biophysics of Cryopreservation

A clear understanding of biophysical behaviour of cells at the time of freezing and exposure to different cryoprotectants is critical in providing conditions to improve the cell structural and functional potential after freezing-thawing. During slow rate of cooling, extracellular ice crystal formation begins with the presence of a nucleation site in the extracellular medium. Because ice is pure crystalline water, the extracellular space becomes hypertonic due to the removal of water as ice crystals develop. Intracellular water, therefore, moves outward across the cell membrane due to the differential osmotic gradient, and cells dehydrate and shrink. This is the opportunity when certain cryoprotective compounds come into play, permeating the cells and protecting them against high solute concentration or ice crystal damage. Because various cryoprotectant agents (CPAs) permeate different cell types at varying rates, it is of benefit to understand the biophysics of cryopreservation to minimize damage.

Freezing Injuries

Two main rival theories have been proposed to explain cell damages due to freezing. One emphasizes the direct and primarily mechanical damage to live cells by ice crystals puncturing through the cell membranes, and the other highlights the secondary effects of ice formation via osmotic changes. Perhaps, both mechanisms are important and what is recently agreed upon is that for individual cells, for example those in suspensions, intracellular freezing is very hazardous, while the extracellular ice may not be as harmful. Unlike cell suspensions, the cellular organization and structural composition of the tissue may be seriously affected by cryogenic damage through widespread extracellular ice formation. Ice formation within a tissue, initiated in the extracellular space, leads to an osmotic gradient across the cell membranes, causing intracellular water to move toward the concentrated extracellular space surrounding the cells. Due to the differential destructive effects of extracellular ice formation between cell suspensions and complex tissues, conventional approaches to cryopreservation of cells, even testis cells for instance, may not necessarily be suitable for multicellular tissues such as the testis tissue. Optimal cooling rates for various cell and tissue types have been shown to differ and be directly associated with the degree of water permeability of cell membranes at different temperatures during freezing.

When extracellular ice formation causes elevated solvent concentrations, it leads to cell dehydration; prolonged exposure to which can permanently damage cell membranes and destabilize proteins. However, short exposure of cells to optimized concentrations of hypertonic media before freezing might protect them from retention of supercooled water within cells and subsequent crystallization during freezing. When cooling is faster than optimal, intracellular ice formation could occur due to inadequate time for water to follow the osmotic gradient across the cell membrane. The osmotic tolerance of cells is another critical factor to be considered during addition and removal of different cryoprotectants. Physical destruction, subsequent organelle disruption, and functional damage are some of the known consequences of ice crystal formation.

Protection Mechanism and Toxicity of Cryoprotectants

Sufficient concentration of cryoprotectants could minimize ice crystallization and/or promote amorphous solidification (vitrification). Glycerol was introduced as a CPA in 1949 and, a decade later, cryoprotective properties of dimethyl sulfoxide (DMSO) were also reported. These two cryoprotectants have mainly been used since then as classic cryoprotective additives, although many other CPAs have been introduced. Permeating CPAs, such as DMSO, glycerol, methanol, propanediol, ethylene glycol, and dimethyl acetaldehyde, as well as non-permeating CPAs, including sucrose, dextran, albumin, polyvinyl pyrollidone, and hydroxyethyl starch, have also been shown to afford effective cryoprotection.

Cryoprotective agents are known to act through different pathways to protect cells against freezing injuries. This includes modulation of hydrogen bonding and interaction with water molecules, which give CPAs solubility and high permeability across cell membranes. As a second mechanism, CPAs may provide a salt-buffering effect. During freezing, cells experience osmotic dehydration and shrinkage; therefore, the addition of CPAs into the cells maintains salt dilution. Basically, the CPA replaces water in cells, which dilutes the intracellular salts and prevents intracellular crystal formation. The amount of CPAs and water that permeates into the cells depends on the concentration of permeable solutes and the final cell volume. The properties of CPAs and those of cell membranes will influence the degree of cryoprotection for different cell types. A third potential pathway is the stabilization of biomembrane critical macromolecules. Under normal conditions, water stabilizes the membrane bilayers. Loss of water during cryopreservation may disrupt normal membrane permeability and damage the membrane itself. The CPAs stabilize proteins as well as phospholipid bilayers of cell membranes and help to protect the membrane against freezing and dehydration stresses. Studies have collectively demonstrated that CPAs, including DMSO and disaccharide sugars such as sucrose and trehalose, may electrostatically interact with membrane phospholipids to provide stabilization. The fourth mechanism by which CPAs protect the cells and tissue is through scavenging oxygen free radicals and preventing oxidative stress to the cells. CPAs block the action of unstable intermediate products, such as oxygen free radicals, by binding their hydrogen atoms to them. The fifth possible pathway for the protective effects of CPAs is the inhibition of nucleation, through which ice formation occurs in the media. During cooling, initial heterogeneous nucleation sites, such as small particles, change in shape and increase in size within media, eventually reaching a stage that forms ice crystals. Alternatively, induced nucleation could be beneficial to provide consistent extracellular crystallization. This phenomenon is the basis for "seeding", which induces nucleation onto supercooled media enabling proper cryopreservation. Seeding can be achieved by clamping the side of vials or straws with a forceps cooled in liquid nitrogen to stimulate local ice growth in the solutions. Intracellular nucleation can also be lethal or damaging for cells and tissues. Some CPAs, such as DMSO or glycerol, inhibit nucleation by increasing the high viscosity of intracellular water. Nonpermeating CPAs, on the other hand, increase and promote cellular duration by increasing the extracellular solute concentration thereby reducing intracellular crystallization.

Despite the protective potential of CPAs, a side effect of their addition is cytotoxicity. Tissue tolerance to CPAs is limited and overexposure may cause damage; however, measuring this toxicity is difficult to precisely assess. Cytotoxicity is further exacerbated by increasing CPA concentrations during ice formation. Optimizing the freezing rate as well as the addition or removal of CPAs could reduce their toxicity.

Choice of Cryopreservation Strategies

For cryopreservation of testicular tissue, two popular strategies are slow freezing and vitrification. These techniques differ mainly in the concentration of CPAs used.

Cryopreservation of cells within intact tissues is obviously more demanding than for cells within suspensions. Theoretical differences include heterogeneity of cells, slower rates of solute diffusion, and heat exchange through the mass of a complex tissue. However, judging from evidence from other tissue types, if a sufficient concentration of CPAs is provided, finding a proper cooling rate can yield high survival for different cell types within the tissue. Critical factors for effective cryopreservation, such as cell permeability to water or CPA and subsequent osmotic changes, are directly affected by the rate of cooling (Mazur, 1990). Therefore, finding the optimal cryopreservation protocol for testicular tissue of a particular species/maturational state depends on the application of a proper concentration of the cryoprotectant with a suitable cooling rate.

Slow (controlled) freezing is considered the conventional method for cryopreservation of testicular tissue, in which the CPA is used at low concentrations (usually 0.5 to 2 M) to minimize both cell damage and CPA toxicity. During slow freezing (e.g., -1 °C/min), the CPA is given a chance to slow down the formation of extracellular ice crystals (and prevent the intracellular ones) but especially to moderate the indirect solution effects as freezing proceeds. However, prolonged exposure to CPA before completion of cryopreservation can also cause cell toxicity. On the other hand, if the cell is cooled more rapidly, then water will not leave the cells fast enough to avoid intracellular freezing, which is very damaging to the cells. Using automated systems, freezing curves can be customized to maximize cell viability after cryopreservation of the tissue.

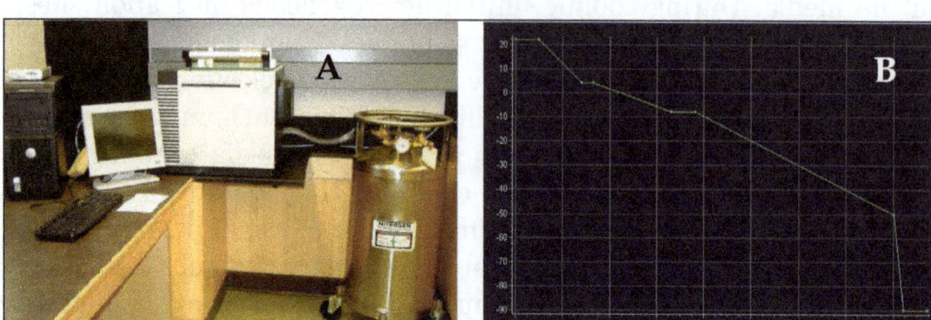

A programmable automated freezing system. Although requiring larger capital investments, automated cell/tissue freezing systems (A), consisting of a freezing chamber attached to a computer and a liquid nitrogen tank, allow customization of the freezing curve (B) to achieve pre-defined temperatures (Y-axis) for desired lengths of time (X-axis), in an accurate and consistent manner.

As indicated earlier, the formation of extracellular ice, which may not pose a problem for freezing of cell suspensions, is likely the main problem for tissues. Therefore, an alternative route to avoid ice crystal formation and solute damage within the tissue is to avoid ice crystal formation altogether using transformation of aqueous milieu of the cell/tissue to the amorphous character of a glassy state, known as vitrification. Vitrification is a cryopreservation method in which ice crystal formation is prevented because the cells or tissues are exposed to very high concentrations of CPAs (e.g., 5 to 8 M) and undergo ultra rapid freezing rates (e.g., up to -2500 °C/min). However, this approach is compromised by the cytotoxic effects of CPAs at such high concentration, especially with increased exposure times. For small volumes of cell suspension, CPA concentrations can be reduced somewhat by using very rapid cooling and warming rates. However, especially with increasing size and complexity of the tissue, the limits of temperature exchange rates are more restricted, hence the use of very high concentrations of CPAs are unavoidable. To overcome this problem, the use of a combination of CPAs to improve vitrification while reducing toxicity has been suggested. Proper media may include disaccharides, such as sucrose or trehalose, and proteins or polymers. The optimal CPA concentrations and exposure times to prevent toxicity must be specifically considered for each tissue type. We have used a solid-surface vitrification method to minimize the volume surrounding the tissue pieces, while avoiding liquid nitrogen (LN2) vapour formation and preventing direct contact with LN2 to prevent potential contamination.

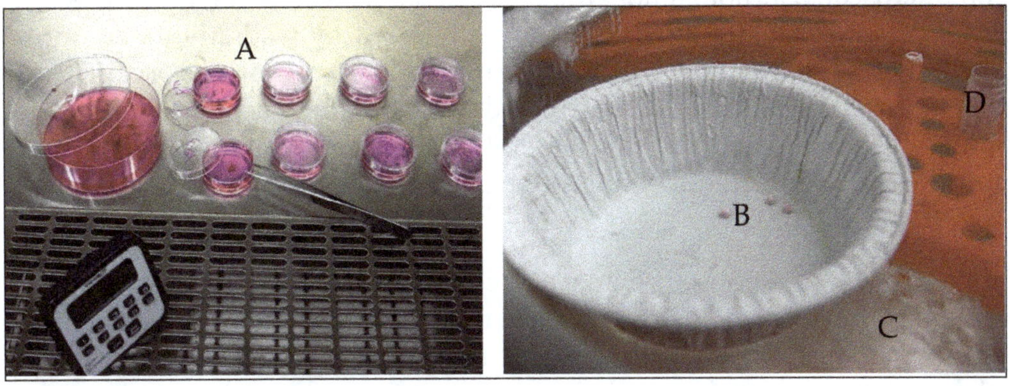

Solid-surface vitrification procedure for testicular tissue fragments. After exposure of testis tissue fragments to differing concentrations of vitrification solutions for varying lengths of time (A), testis tissue fragments are placed on a sterile aluminum boat (B) floating on liquid nitrogen (C), then transferred into cooled cryovials (D) followed by plunging into liquid nitrogen.

Thawing Methods

Whether freezing is permitted (conventional cryopreservation) or prevented (vitrification), the CPA that has reached the internal compartments of a multicellular system must diffuse back through numerous membranes in the tissue, with each acting as a barrier. Therefore, optimal thawing and CPA removal procedures are also critical factors for cell/tissue survival after freezing. Earlier studies pointed out that consistent cooling and thawing rates (slow-freezing followed by slow-thawing, or fast-freezing followed by fast-thawing) can improve cell/tissue survival after cryopreservation. Moreover, extreme osmotic changes during CPA removal might damage the cells by extensive cell shrinkage or swelling associated with the rapid movement of water into the cell as compared to the slower movement of the CPA out of the cell. However, a limited amount of water replacement is needed to restore osmotic equilibrium and physiologic cell volume.

Post-thawing Analysis

For successful cryopreservation of a complex vascularized tissue, such as testis tissue, the majority of essential cells need to be viable for the tissue to survive and retain its function. However, there is not yet a comprehensive and universally applied method for postthawing analysis of cryopreserved testis tissue; subsequently, multiple approaches have been used to assess tissue/cell viability and extent of cryogenic injuries. These approaches commonly include histopathological examination of tissue sections for morphological changes. Using light microscopy, for instance, such objective criteria as seminiferous cord/tubular diameter or cell density within tubule cross sections can be measured, or semiquantitative morphometric analyses applied to subjectively score such criteria as health or integrity of tissue compartments. Transmitted electron microscopy, although not widely used, can be invaluable in the examination of subcellular components most likely to be affected by testis tissue cryopreservation, including cytoplasm integrity, nuclear membrane, and various organelles. Other valuable morphological analyses may include assessment of cell-specific changes, for example, using double-staining of proliferation markers (e.g., Ki67) and MAGE-AH, vimentin, or CD34 for identification of spermatogonia, Sertoli cells, or peritubular cells, respectively.

A quantitative measure of tissue damage due to cytotoxicity after cryopreservation can be achieved through lactate dehydrogenase release assays or through viability assessment of dissociated cells after digestion of frozen-thawed tissues using Trypan blue exclusion assays or the various cell viability kits using a flow cytometer analyzer. Assessment of apoptosis, using for instance, caspase-3, or TUNEL assay for detection of DNA fragmentation provides insight into the extent of cell damage. Detection of phophatidylserine translocation from the inner to the outer layer of the plasma membrane, using fluorescent-labelled Annexin V, also allows more targeted assessment of apoptotic associated changes within the cryopreserved testis tissue.

Having merely high cell survival rates or lacking visible damage does not guarantee functional preservation of the tissue as a whole. A thorough post-thawing analysis should include a form of testing for the functionality of the cryopreserved tissue. Post-thawing in vitro organotypic culture of the cryopreserved testis tissue has allowed assessment of its survival in the shortterm and measurement of its hormone release into culture media. Perhaps more robust examination is provided by grafting, where the survival and developmental competence (both in terms of germ cell differentiation and androgen release) of the cryopreserved tissue in vivo as grafts allows a longer-term functional assessment.

Effects of Tissue Size

To offer cryoprotection, the CPAs need to diffuse rapidly in and out of the tissue; therefore, the size of testis tissue samples undergoing cryopreservation can be an important intuitive consideration. The results of studies differ depending not only with respect to the donor species but also potentially on the protocols employed. For instance, while cryopreservation of immature rat testis using similar procedures demonstrated better results for 7.5 mg pieces than 15 mg, cryopreservation of immature mouse testis using whole testes with punctured tunica albuginea was deemed more suitable than using whole testes with intact tunica, whole testes without tunica, or testis halves. Mouse testes have considerably less connective tissue content than most other species; therefore, tissue fragment size is especially a concern for testis tissues from species with higher interstitial tissue density. For cryopreservation of (cryptorchid) testes from prepubertal boys, fragments sizes of 2-9 mm3 were used successfully. We also reported that immature porcine testis tissues undergoing the same cryopreservation treatments were not affected by the original size of the testis tissue fragment (5, 15, 20, or 30 mg). Although not used for cryopreservation, no effect of tissue sample size was observed for one-wk old piglet testes (as intact or fragments of 100 or 30 mg) when used for hypothermic preservation for 6 days. It remains to be seen if whole human testes can be cryopreserved as has been accomplished for whole ovaries.

Applications of Testis Cryopreservation for new Reproductive Technologies

Given that properly cryopreserved testis biopsies can last decades in liquid nitrogen and that most prepubertal cancer patient boys donating biopsies may not need to resort to assisted reproductive technologies for a couple of decades, it is advisable that cryopreservation of testicular biopsies be offered to such patients in a hope that our ability to use such tissues will be further improved and the options expanded in the coming years.

A number of potential applications already exist for the use of cryopreserved testicular tissue in experimental and clinical settings in reproduction medicine/science. Such technologies allow retrieval of existing sperm from mature donor samples and, more importantly, offer hope for production of sperm in samples of cryopreserved testis immature testis. If the preserved testis tissue contains endogenous spermatogenesis (e.g., from obstructive azoospermic adult patients), it can be used to extract sperm, elongated spermatids, or even round spermatids to be used for fertilization of oocytes through ICSI.

If preserved testis samples are obtained from neonatal/immature donors, they can still be used to induce spermatogenesis through the following approaches.

Germ Cell Transplantation

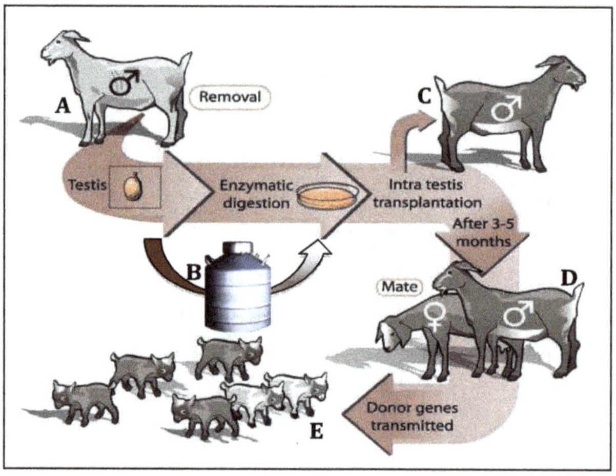

Schematic overview of germ cell transplantation from a donor male into the testes of a recipient. The testes are collected from a donor animal (A), which could theoretically include post-mortem testis recovery from a recently deceased juvenile individual of an endangered species. The testis tissue could be cryopreserved (B) until conditions for its use are in place. At the time of transplantation, a single-cell suspension is prepared and the cells are infused into the seminiferous tubules of a recipient animal (C). Mating of the recipient (D) produces progeny (E), some of which will carry the donor genome.

The technique for germ cell transplantation has allowed (re)establishment of spermatogenesis after introduction of donor testis cell suspensions into the seminiferous tubules of infertile recipient testes. Once deposited in the tubular lumen, donor SSCs are recognized by the host Sertoli cells and allowed passage to the stem cell niche, where new colonies of spermatogenesis can begin and expand. This approach has allowed production of sufficient numbers of sperm to allow infertile recipient mice to sire donor-derived progeny. Later, the capability of cryopreserved mouse testis cells after transplantation into recipient testes to start spermatogenesis was also confirmed. While heterologous transplantation of human germ cells into recipient mice did not lead to completion of spermatogenesis, the transfer technique has been tested using human testes. Although autologous/homologous transplantation of germ cells for humans is currently considered purely experimental, one possibility for prepubertal human testis samples taken and frozen prior to treatments is to isolate testis cells and transfer them back to the individual. As a major problem with this approach is the risk of reseeding a systemic cancer, solutions to this (e.g., soring out tumour cells) and other safety issues are under investigation.

We have expanded the technique for germ cell transplantation into farm animals, showed the feasibility of SSC engraftment in unrelated recipient individuals (of the same species) without a need for immune-suppression, and further demonstrated the applicability of the approach through donor-derived sperm production by the recipients and birth of progeny carrying the donor characteristics. Therefore, although experimental at this stage, the approach may offer promise in salvaging genetic material from cryopreserved testicular tissue from immature endangered species.

Testis Tissue (Xeno)grafting

Another potential strategy for the use of cryopreserved testis tissue is represented by testis tissue xenografting. Grafting of both fresh and cryopreserved testis tissue fragments from donors of different species under the back skin of recipient mice results in the production of functional sperm. The approach has especially been successful using neonatal/immature donors from laboratory animals to domestic animals, primates, and even humans.

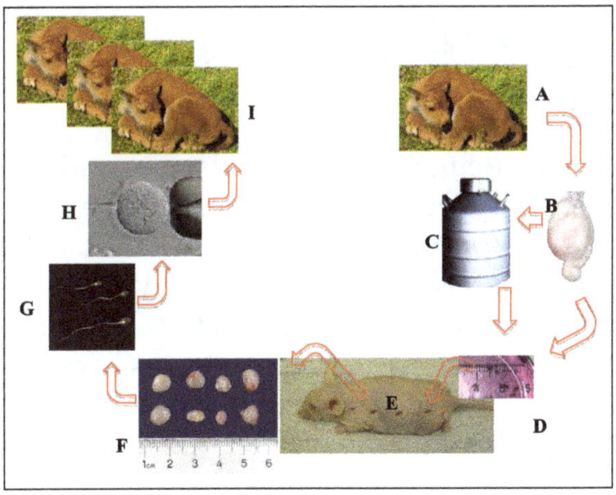

Schematic representation of testis tissue (xeno)grafting from an immature donor individual into the back skin of a host mouse. The testes are collected from a donor animal (A), which could include post-mortem testis recovery from a recently deceased newborn animal of an endangered species. The testis tissue (B) could be cryopreserved (C) until grafting. At the time of grafting, tissue fragments of ~0.5 mm3 (D) are prepared and the fragments are grafted subcutaneously into an immunodeficient host mouse (E). When given enough time, the grafts can grow in size (F) and undergo development, leading to the production of complete spermatogenesis, including fertilization-competent sperm (G). The sperm can then be extracted from the grafts and used in intracytoplasmic sperm injection (ICSI) (H), which after embryo transfer can potentially lead to birth of progeny (I).

The sperm recovered from such grafts, including those from primates, have been shown to be fertilization competent after ICSI, leading to the birth of healthy progeny. We recently showed that testes recovered post-mortem from newborn bison calves, as a model for closely-related rare or endangered ungulates can be used for this application, and when allowed to develop in the host mouse, lead to full spermatogenesis. Therefore, testis tissue xenografting can be used as unique solution for genetic conservation of immature males by producing sperm from these otherwise resource-less donors in xenografts, followed by extraction and cryopreservation of sperm for future use in ICSI.

However, xenografting of human gonadal tissues into animals to harvest the resultant gametes for use in IVF for humans is prohibited in Canada, and possibly in other countries, due to the potentially serious risk of animal viral transmission or contamination with animal genetics. Nevertheless, the promising results from animal research suggest a potential hope for future use of cryopreserved testis biopsies from pre-adolescent boys to be grafted back to the individual; whether this technique can be used to produce viable sperm for future use from prepubertal boys undergoing gonadotoxic treatments remains to be determined. However, the same safety risks as for autologous germ cell transplantation exist and require addressing before such an option can be offered clinically.

In Vitro Maturation of Germ Cells

In theory, cryopreserved testicular tissues can also be used for in vitro induction of differentiated germ cells and ideally production of sperm or spermatids to be used for ICSI. If successful, this approach can circumvent the potential risk of reintroducing cancer cells into post-recovery patients. Many labs have experimented with the idea, and some have had success with maturation of later stages of human spermatogenesis (but not from SSCs), including live births. Availability of a culture system to support complete in vitro spermatogenesis from the SSC stage was, however, elusive until very recently when it was reported that all spermatogenic lineage cells including fertilization competent sperm could be produced from neonatal mouse testes maintained exclusively in a culture system. This is a very promising step, indicating that similar results may be achievable in future using immature human testis biopsies.

Current Trends in Testis Tissue Cryopreservation

Since the first reports of successful germ cell transplantation and xenografting of testis tissue raised new interest in this field, several promising cryopreservation protocols have been introduced. Perhaps not surprisingly, the results differed and at times conflicted depending on the tissue donor species/developmental stage. These first reports of cryopreservation of pig and mouse testis tissues were based on DMSO-based slow freezing protocols originally developed for isolated testis cells or for ovarian tissue, respectively. Later, other detailed studies comparing multiple protocols showed high cell viability with programmed slow-freezing of immature mouse testis tissue using 1.5M DMSO as a cryoprotectant. DMSO has also been found to be a more suitable cryoprotective agent than ethylene glycol for immature mouse and rat testis tissue. Shinohara et al. reported the birth of mouse offspring from sperm retrieved from cryopreserved pre-pubertal testis tissue with DMSO after transplantation under tunica albuginea of the recipient testes. Similar results were obtained using primate testis tissue, where 1.4M (but not 0.7M) DMSO was able to protect some of the developmental potential of grafts from rhesus monkeys but the 0.7M DMSO protocol was successful for cryopreservation of human testis tissue at one age/developmental stage but not others. Somewhat different from reports in other species, and after an extensive study of several strategies for cryopreservation of immature testis tissue, we concluded that glycerol was a better cryoprotectant for pig tissues. These results suggest that each species and donor developmental age may need a different cryopreservation protocol, with a concomitant need to adjust the concentration of cryoprotectant or even adopt different cryoprotectants. These differences may be related to testicular architecture, morphology, or lipid composition.

In a first report of immature testis tissue vitrification, we also showed maintenance of cell viability and developmental potential to actively (re)establish complete spermatogenesis after xenografting into immunodeficient mice. Recently, similar or much higher cell viability results were obtained using immature mouse testis tissue with vitrification compared with conventional slow freezing. With proper tissue handling, and the use of an appropriate choice of final cryoprotectant exposure, vitrification can provide preferential conditions for tissue freezing with proven superior results in restoration of immature testis tissue. Vitrification also does not require the extensive laboratory equipment commonly used for programmed slow freezing; however, direct plunging of tissues into liquid nitrogen, a common procedure in routine vitrification, poses a greater risk of contamination. The solid-surface vitrification of testis tissue is an easy, safe, and applicable cryopreservation technique for the preservation of tissue structural integrity and developmental potential.

SEMEN CRYOPRESERVATION

Semen cryopreservation (commonly called sperm banking or sperm freezing) is a procedure to preserve sperm cells. Semen can be used successfully indefinitely after cryopreservation. For human sperm, the longest reported successful storage is 24 years. It can be used for sperm donation where the recipient wants the treatment in a different time or place, or as a means of preserving fertility for men undergoing vasectomy or treatments that may compromise their fertility, such as chemotherapy, radiation therapy or surgery.

Freezing

The most common cryoprotectant used for semen is glycerol (10% in culture medium). Often sucrose or other di-, trisaccharides are added to glycerol solution. Cryoprotectant media may be supplemented with either egg yolk or soy lecithin, with the two having no statistically significant differences compared to each other regarding motility, morphology, ability to bind to hyaluronate in vitro, or DNA integrity after thawing.

Additional cryoprotectants can be used to increase sperm viability and fertility rates post-freezing. Treatment of sperm with heparin binding proteins prior to cryopreservation showed decreased cryoinjury and generation of ROS. The addition of nerve growth factor as a cryoprotectant decreases sperm cell death rates and increased motility after thawing. Incorporation of cholesterol into sperm cell membranes with the use of cyclodextrins prior to freezing also increases sperm viability.

Semen is frozen using either a controlled-rate, slow-cooling method (slow programmable freezing or SPF) or a newer flash-freezing process known as vitrification. Vitrification gives superior post-thaw motility and cryosurvival than *slow programmable freezing*.

Thawing

Thawing at 40 °C seems to result in optimal sperm motility. On the other hand, the exact thawing temperature seems to have only minor effect on sperm viability, acrosomal status, ATP content, and DNA. As with freezing, various techniques have been developed for the thawing process.

Refreezing

In terms of the level of sperm DNA fragmentation, up to three cycles of freezing and thawing can be performed without causing a level of risk significantly higher than following a single cycle of freezing and thawing. This is provided that samples are refrozen in their original cryoprotectant and are not going through sperm washing or other alteration in between, and provided that they are separated by density gradient centrifugation or swim-up before use in assisted reproduction technology.

Effect on Quality

Some evidence suggests an increase in single-strand breaks, condensation and fragmentation of DNA in sperm after cryopreservation. This can potentially increase the risk of mutations in offspring DNA. Antioxidants and the use of well-controlled cooling regimes could potentially improve outcomes.

In long-term follow-up studies, no evidence has been found either of an increase in birth defects or chromosomal abnormalities in people conceived from cryopreserved sperm compared with the general population.

OOCYTE CRYOPRESERVATION

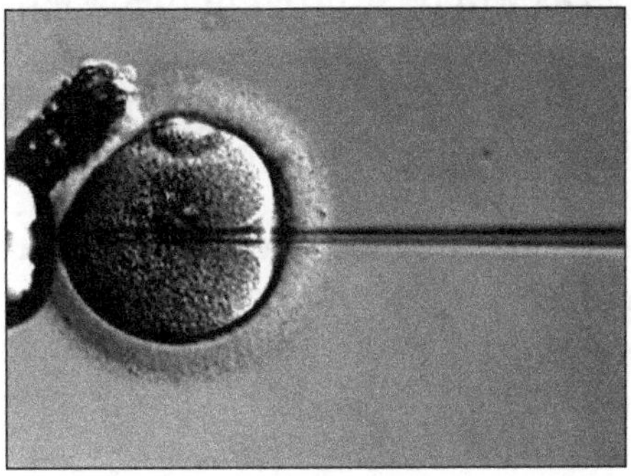

ICSI sperm injection into oocyte.

Human oocyte cryopreservation (egg freezing) is a procedure to preserve a woman's eggs (oocytes). This technique has been used to enable women to postpone pregnancy to a later date - whether for medical reasons such as cancer treatment or for social reasons such as employment or studying. Several studies have proven that most infertility problems are due to germ cell deterioration related to aging. Surprisingly, the uterus remains completely functional in most elderly women. This implies that the factor which needs to be preserved is the woman's eggs. The eggs are extracted, frozen and stored. The intention of the procedure is that the woman may choose to have the eggs thawed, fertilized, and transferred to the uterus as embryos to facilitate a pregnancy in the future. The procedure's success rate (the chances of a live birth using frozen eggs) varies depending on the age of the woman, and ranges from 14.8 percent (if the eggs were extracted when the woman was 40) to 31.5 percent (if the eggs were extracted when the woman was 25).

Indications

Oocyte cryopreservation can increase the chance of a future pregnancy for three key groups of women:

- Those diagnosed with cancer who have not yet begun chemotherapy or radiotherapy,
- Those undergoing treatment with assisted reproductive technologies who do not consider embryo freezing an option,
- Those who would like to preserve their future ability to have children, either because they do not yet have a partner, or for other personal or medical reasons.

Over 50,000 reproductive-age women are diagnosed with cancer each year in the United States. Chemotherapy and radiotherapy are toxic for oocytes, leaving few, if any, viable eggs. Egg freezing offers women with cancer the chance to preserve their eggs so that they can attempt to have children in the future.

Oocyte cryopreservation is an option for individuals undergoing IVF who object, either for religious or ethical reasons, to the practice of freezing embryos. Having the option to fertilize only as many eggs as will be utilized in the IVF process, and then freeze any remaining unfertilized eggs can be a solution. In this way, there are no excess embryos created, and there need be no disposition of unused frozen embryos, a practice which can create complex choices for certain individuals.

Additionally, women with a family history of early menopause have an interest in fertility preservation. With egg freezing, they will have a frozen store of eggs, in the likelihood that their eggs are depleted at an early age.

Method

The egg retrieval process for oocyte cryopreservation is the same as that for in vitro fertilization. This includes one to several weeks of hormone injections that stimulate ovaries to ripen multiple eggs. When the eggs are mature, final maturation induction is performed, preferably by using a GnRH agonist rather than human chorionic gonadotrophin (hCG), since it decreases the risk of ovarian hyperstimulation syndrome with no evidence of a difference in live birth rate (in contrast to fresh cycles where usage of GnRH agonist has a lower live birth rate). The eggs are subsequently removed from the body by transvaginal oocyte retrieval. The procedure is usually conducted under sedation. The eggs are immediately frozen.

The egg is the largest cell in the human body and contains a high amount of water. When the egg is frozen, the ice crystals that form can destroy the integrity of the cell. To prevent this, the egg must be dehydrated prior to freezing. This is done using cryoprotectants which replace most of the water within the cell and inhibit the formation of ice crystals.

Eggs (oocytes) are frozen using either a controlled-rate, slow-cooling method or a newer flash-freezing process known as vitrification. Vitrification is much faster but requires higher concentrations of cryoprotectants to be added. The result of vitrification is a solid glass-like cell, free of ice crystals. Indeed, freezing is a phase transition. Vitrification, as opposed to freezing, is a physical transition. Realizing this fundamental difference, vitrification concept has been developed and successfully applied in IVF treatment with the first life birth following vitrification of oocytes achieved in 1999. Vitrification eliminates ice formation inside and outside of oocytes on cooling, during cryostorage and on warming. Vitrification is associated with higher survival rates and better development compared to slow-cooling when applied to oocytes in metaphase II (MII). Vitrification has also become the method of choice for pronuclear oocytes, although prospective randomized controlled trials are still lacking.

During the freezing process, the zona pellucida, or shell of the egg can be modified preventing fertilization. Thus, currently, when eggs are thawed, a special fertilization procedure is performed by an embryologist whereby sperm is injected directly into the egg with a needle rather than allowing sperm to penetrate naturally by placing it around the egg in a dish. This injection technique is called ICSI (Intracytoplasmic Sperm Injection) and is also used in IVF.

Immature oocytes have been grown until maturation *in vitro*, but it is not yet clinically available.

Success Rates

The percentage of transferred cycles is lower in frozen cycles compared with fresh cycles (approx. 30% and 50%). Such outcomes are considered comparable.

In a 2013 meta-analysis of more than 2,200 cycles using frozen eggs, scientists found the probability of having a live birth after three cycles was 31.5 percent for women who froze their eggs at age 25, 25.9 percent at age 30, 19.3 percent at age 35, and 14.8 percent at age 40.

Two recent studies showed that the rate of birth defects and chromosomal defects when using cryopreserved oocytes is consistent with that of natural conception.

Recent modifications in protocol regarding cryoprotectant composition, temperature and storage methods have had a large impact on the technology, and while it is still considered an experimental procedure, it is quickly becoming an option for women. Slow freezing traditionally has been the most commonly used method to cryopreserve oocytes, and is the method that has resulted in the most babies born from frozen oocytes worldwide. Ultra-rapid freezing or vitrification represents a potential alternative freezing method.

In the fall of 2009, The American Society for Reproductive Medicine (ASRM) issued an opinion on oocyte cryopreservation concluding that the science holds "great promise for applications in oocyte donation and fertility preservation" because recent laboratory modifications have resulted in improved oocyte survival, fertilization, and pregnancy rates from frozen-thawed oocytes in IVF. The ASRM noted that from the limited research performed to date, there does not appear to be an increase in chromosomal abnormalities, birth defects, or developmental deficits in the children born from cryopreserved oocytes. The ASRM recommended that, pending further research, oocyte cryopreservation should be introduced into clinical practice on an investigational basis and under the guidance of an Institutional Review Board (IRB). As with any new technology, safety and efficacy must be evaluated and demonstrated through continued research.

In October 2012, the ASRM lifted the experimental label from the technology for women with a medical need, citing success rates in live births, among other findings. However, they also warned against using it only to delay child-bearing.

In 2014, a Cochrane systematic review about this topic was published. It compared vitrification (the newest technology) versus slow freezing (the oldest one). Key results of that review showed that the clinical pregnancy rate was almost 4 times higher in the oocyte vitrification group than in the slow freezing group, with moderate quality of evidence.

Immature oocytes have been grown until maturation *in vitro* at a 10% survival rate, but no experiment has been performed to fertilize such oocytes.

Cost

The cost of the egg freezing procedure (without embryo transfer) in the United States, the United Kingdom and other European countries varies in between $5,000 and $12,000. This does not include the fertility medications involved in the procedure which can cost between $4,000 and $5,000. The cost of egg storage can vary from $100 to more than $1,000. It is important for women to be aware that provisional health programs do not cover social egg freezing. Furthermore, no provinces provide funding for IVF after social egg freezing.

Medical tourism may have lower costs than performing egg freezing in high-cost countries like the US. Some well established medical tourism and IVF countries such as Czech Republic, Ukraine and Cyprus offer egg freezing at competitive prices. It is a lower cost alternative to typical US options for egg freezing. Spain and the Czech Republic are popular destinations for this treatment.

Elective Oocyte Cryopreservation

Elective oocyte cryopreservation, also known as social egg freezing, is non-essential egg freezing for the purpose of preserving fertility for delayed child-bearing when natural conception becomes more problematic. The frequency of this procedure has steadily increased since October 2012 when the American Society for Reproductive Medicine (ASRM) lifted the 'experimental' label from the process. There was a spike in interest in 2014 when global corporations Apple and Facebook revealed they were introducing egg freezing as a benefit for their female employees. This announcement was controversial as some women found it empowering and practical, while others viewed the message these companies were sending to women trying to have a successful long-term career and a family as harmful and alienating. A string of "egg freezing parties" hosted by third-party companies have also helped popularize the concept among young women. Social science research suggests that women use elective egg freezing to disentangle their search for a romantic partner from their plans to have children.

In 2016, then US Secretary of Defense Ash Carter announced that the Department of Defense will cover the cost of freezing sperm or eggs through a pilot program for active duty service members, with the intention of preserving their ability to start a family even if they suffer certain combat injuries.

There are still warnings for women using this technology to fall pregnant at an older age as the risk of pregnancy complications increases with a mother's age. However, studies have shown that the risk of congenital abnormalities in babies born from frozen oocytes is not increased further when compared to naturally conceived babies.

Risks

The risks for women undergoing egg freezing can include: vaginal/uterine bleeding from the oocyte recovery procedure; developing ovarian hyperstimulation syndrome (OHSS) as a reaction to the hormones used to induce hyperovulation (producing more than one egg); and liver failure. There are indications that ovarian stimulation may increase risk of breast, uterine and other cancers, however this remains inconclusive. The long-term effects of egg extraction on women's bodies have not been well studied.

There may also be some risks to any resulting child. These include IVF-associated risks such as multiple pregnancy, pregnancy-induced high blood pressure, premature delivery, operative delivery and infants displaying low birth weight. There also remain potential unknown risks caused by long-term freezing. Expanded use of Intracytoplasmic sperm injection (ICSI) to inject a single sperm into a thawed egg could additionally be a cause for concern as this method has been associated with a higher rate of birth defects.

OVARIAN TISSUE CRYOPRESERVATION

Ovarian tissue cryopreservation is cryopreservation of tissue of the ovary of a female.

Indications

Cryopreservation of ovarian tissue is of interest to women who want fertility preservation beyond the natural limit, or whose reproductive potential is threatened by cancer therapy, for example in hematologic malignancies or breast cancer. It can be performed on prepubertal girls at risk for premature ovarian failure, and this procedure is as feasible and safe as comparable operative procedures in children.

Procedure

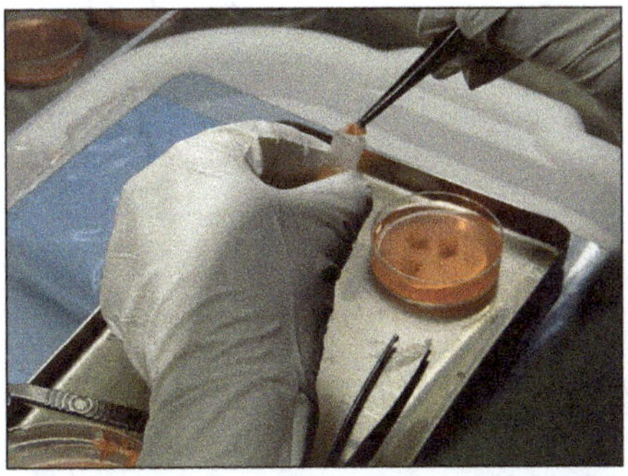

Placing ovarian tissue strips into the preserving solution.

Cryopreserving ovarian tissue strips.

The procedure is to take a part of the ovary and carry out slow freezing before storing it in liquid nitrogen whilst therapy is undertaken. Tissue can then be thawed and implanted near the fallopian, either orthotopic (on the natural location) or heterotopic (on the abdominal wall), where it starts to produce new eggs, allowing normal conception to take place. A study of 60 procedures concluded that ovarian tissue harvesting appears to be safe. A study has also concluded that culturing a thawed fetal ovarian tissue for a few days before transplanting can be beneficial to the development of follicles.

Strips of cortical ovarian tissue can also be cryopreserved, but it must be re-implanted into the body to allow the encapsulated immature follicles to complete their maturation. In vitro maturation has been performed experimentally, but the technique is not yet clinically available. With this technique, cryopreserved ovarian tissue could possibly be used to make oocytes that can directly undergo in vitro fertilization.

Risk of Cancer Recurence

For autotransplantation of cryopreserved ovarian tissue in cancer survivors, metastases have been repeatedly detected in ovarian tissue obtained from patients with leukemia, as well as in one patient with Ewing sarcoma. Ovarian tissue autotransplantation may pose a risk of cancer recurrence in patients with colorectal, gastric and endometrial cancer. However, no metastases have been detected in ovarian tissue from lymphoma and breast cancer patients who have been undergoing ovarian tissue cryopreseration.

EMBRYO CRYOPRESERVATION

Cryopreservation of embryos is the process of preserving an embryo at sub-zero temperatures, generally at an embryogenesis stage corresponding to pre-implantation, that is, from fertilisation to the blastocyst stage.

Indications

Embryo cryopreservation is useful for leftover embryos after a cycle of in vitro fertilisation, as patients who fail to conceive may become pregnant using such embryos without having to go through a full IVF cycle. Or, if pregnancy occurred, they could return later for another pregnancy. Spare oocytes or embryos resulting from fertility treatments may be used for oocyte donation or embryo donation to another woman or couple, and embryos may be created, frozen and stored specifically for transfer and donation by using donor eggs and sperm.

Method

Embryo cryopreservation is generally performed as a component of in vitro fertilization (which generally also includes ovarian hyperstimulation, egg retrieval and embryo transfer). The ovarian hyperstimulation is preferably done by using a GnRH agonist rather than human chorionic gonadotrophin (hCG) for final oocyte maturation, since it decreases the risk of ovarian hyperstimulation syndrome with no evidence of a difference in live birth rate (in contrast to fresh cycles where usage of GnRH agonist has a lower live birth rate).

The main techniques used for embryo cryopreservation are vitrification versus slow programmable freezing (SPF). Studies indicate that vitrification is superior or equal to SPF in terms of survival and implantation rates. Vitrification appears to result in decreased risk of DNA damage than slow freezing.

Direct Frozen Embryo Transfer: Embryos can be frozen by SPF in ethylene glycol freeze media and transfer directly to recipients immediately after water thawing without laboratory thawing process. The world's first crossbred bovine embryo transfer calf under tropical conditions was produced by such technique on 23 June 1996 by Dr. Binoy S Vettical of Kerala Livestock Development Board, Mattupatti.

Prevalence

World usage data is hard to come by but it was reported in a study of 23 countries that almost 42,000 frozen human embryo transfers were performed during 2001 in Europe.

Pregnancy Outcome and Determinants

In current state of the art, early embryos having undergone cryopreservation implant at the same rate as equivalent fresh counterparts. The outcome from using cryopreserved embryos has uniformly been positive with no increase in birth defects or development abnormalities, also between fresh versus frozen eggs used for intracytoplasmic sperm injection (ICSI). In fact, pregnancy rates are increased following frozen embryo transfer, and perinatal outcomes are less affected, compared to embryo transfer in the same cycle as ovarian hyperstimulation was performed. The endometrium is believed to not be optimally prepared for implantation following ovarian hyperstimulation, and therefore frozen embryo transfer avails for a separate cycle to focus on optimizing the chances of successful implantation. Children born from vitrified blastocysts have significantly higher birthweight than those born from non-frozen blastocysts. For early cleavage embryos, frozen ones appear to have at least as good obstetric outcome, measured as preterm birth and low birthweight for children born after cryopreservation as compared with children born after fresh cycles.

Oocyte age, survival proportion, and number of transferred embryos are predictors of pregnancy outcome.

Pregnancies have been reported from embryos stored for 16 years. A study of more than 11,000 cryopreserved human embryos showed no significant effect of storage time on post-thaw survival for IVF or oocyte donation cycles, or for embryos frozen at the pronuclear or cleavage stages. In addition, the duration of storage had no significant effect on clinical pregnancy, miscarriage, implantation, or live birth rate, whether from IVF or oocyte donation cycles.

A study in France between 1999 and 2011 came to the result that embryo freezing before administration of gonadotoxic chemotherapy agents to females caused a delay of treatment in 34% of cases, and a live birth in 27% of surviving cases who wanted to become pregnant, with the follow-up time varying between 1 and 13 years. From 1 October 2009 human embryos are allowed to be stored for 10 years in the UK, according to the Human Fertilisation and Embryology Act 2008.

CRYOPRESERVATION METHODS

The viability and extend of cryoinjury in a cell depend on the amount of water in it and the ability of that water to crystallize during the freezing process. The variables that influence the survival rate of a cell are the concentration of CPAs, the cooling rate, the storage temperature and the rewarming rate used. The effects of cooling rates, the use of cryoprotective agents and their mechanisms of action have been extensively described in the previous section. In this part the significance of the storage temperature and the current equipment system technologies are outlined.

Initial Freezing

All cryopreservation processes, despite the long-term storage method used afterwards, begin with the initial freezing of the sample, during which the sample is transferred from its physiological environment to a cryoprotective solution, to minimise the cryoinjury effects. This phase is also known as hypothermic phase and the way it works is by diminishing the cell's metabolism and the chemical toxicity of the CPAs. Biological materials can be introduced into the hypothermic phase either by being directly packaged into the precooled (0 to 4 °C) cryoprotective solution and then transferred to a cryopreservation chamber of the same temperature range, or by being placed into a control rate freezer. However, the later method has been proven to result in better post-cryopreservation survival rates, as it provides robust temperature compensation for the release of the latent heat produced by the cells. This heat release is continuous and caused by the freezing process, as ice formation is an exothermic process.

Equilibrium Freezing

Two are the main methodologies currently used for the initial freezing of cells; equilibrium freezing and vitrification. During equilibrium freezing, cells are exposed to relatively low CPA concentrations and then slowly cooled at a controlled cooling rate, in order to gradually freeze the extracellular solution and in this way to reduce the probability of IIF.

During equilibrium freezing, the cells are exposed to low concentrations of CPAs, equilibrated at room temperature for a short period (10 to 25 min), placed into containers (straws or vials), sealed and stored into a controlled-rate freezer. Thus, the cells are exposed to gradual increase of CPAs concentrations, as more extracellular ice is formed.

To avoid the lethal consequences of IIF and solute concentrations, the controlled freezing systems must facilitate the equilibration of the sample and its surrounding medium, at a temperature marginally below the sample's freezing point. To do that, the sample is forced to freeze in specific regions, by introducing several nucleation points in the cell. This process is called seeding and is induced extracellularly either by contacting the cell with a cold instrument or by a short-lived introduction of liquid nitrogen (LN_2) into the preservation chamber. As a result a solute gradient is created between the intracellular and extracellular medium, which forces the freezing water to flow out of the cell, minimizing that way the chances of IIF.

Equilibrium freezing has three major disadvantages. First of all, the ice crystals formed in the extracellular solution may cause direct mechanical damage to the cell membrane or other fine structures (e.g. sperm tails) and lead to the loss of cell's biophysical function and viability. Secondly, in order to achieve the highest survival rates of the preserved cells, a very tightly controlled optimal cooling rate is required, which many existing cryogenic systems find difficulties to obtain. Finally, the cooling requirements differ from one cell type to another, as the optimal cooling rate depends on individual cell characteristics. As a result, different cell types require different cooling devices, a fact that limits the application of the equilibrium freezing method.

Vitrification

An alternative approach is vitrification, by which any signs of ice formation are prevented. This method maintains the cell as a whole in a vitreous state and precludes both intracellular and extracellular ice formation. In order to achieve that, high concentration of CPAs and a relatively fast cooling rate in excess of the critical cooling rates (the minimum cooling rate to vitrify a solution) are used. During the process, the liquid phase of the cell eventually becomes so viscous that the water molecules do not have enough energy to form crystalline structures, and as a result they start to become amorphous solid or glass. The temperature at which this phenomenon takes place is known as glass transition temperature and below this temperature the cell considers to be biologically inert.

Vitrification with no or low concentrations of CPAs would be suitable for almost all cell types, but it would require the freezing of the samples at ultra-fast cooling rates, on the order of tens of thousands of degrees Celsius per minute. The only method that approaches ultra-fast freezing is the drop of a small cell suspension volume directly into LN_2, but the cooling rate cannot be controlled. A thin vapour coat film forms around the surface of the sample when plunged in LN_2, insulates the sample and prevents a more rapid temperature loss (Leidenfrost effect). Vitrification protocols that require the direct contact of the sample to the LN_2 raise concerns of potential viral contamination, while the whole process of vitrification is very challenging, due to the limited amount of time of cell handling and loading. New techniques of vitrification, though, show very promising results, which overcome the current shortcomings of rapid freezing. For examples, S^3 vitrification exposes the sample to a series of CPA solutions, loaded into sterile straws (0.25 ml), heat sealed them at both ends and vitrifies them. In this way, there is sufficient time for exposure in the CPA solution, loading and vitrification, while the relatively large sized container and the significant lower cooling rates, simplifies the vitrification process and diminishes the amount of LN_2 used.

Laser Pulse Vitrification

In 1998 Fowler and Toner developed a protocol for successful recovery of rapidly frozen red blood cells by vitrification, without the use of CPAs. There technique included the use of a laser pulse prior thawing, which selectively targeted the intracellular ice, so that it was melted and resolidified into a glass phase. The erythrocytes were loaded into straws (1mm x 1μm), plunged into methanol slurry (-96 °C) and transferred into LN_2. The average cooling rate of the samples was approximately 10,000 °C/min. Then, the samples were removed from the LN_2, placed on a cryostage (-170 °C) and a laser pulse (7 ns width, 532 nm wavelength) irradiated the haemoglobin, which exists in the intracellular solution. The short laser pulse was able to be absorbed only by the intracellular medium, which to the increase of its temperature above 0 °C, while the extracellular solution remained at -170 °C. Up to 80% of cells treated that way remained viable after thawing.

Liquidus Tracking (LT)

The principle of LT is the dynamic control of CPA concentrations, throughout the cooling process, in order to maintain the cell just above its freezing point at all times, without the formation of ice. As a result, vitrification can occur without the extremely high CPA concentrations and without the need of ultra-rapid cooling rates. LT systems adjust the CPA concentrations, temperature and time program to the characteristics of the stored cells at all stages of the process. LT systems consist of a conventional controlled rate freezer, with a cooling rate of 0.1 to 0.3 °C/min, while the solution composition is controlled by standard peristaltic pumps.

Isochoric and Hyperbaric Cryopreservation

All the above cryopreservation methods are performed under constant pressure of 1atm of the chamber (isobaric process). A new approach is the freezing under pressure, i.e. the freezing under constant volume of the chamber. Two techniques are currently used, the isochoric and hyperbaric cryopreservation, and both of them use none or few amounts of CPA solutions. The isochoric method has proved to minimise significantly the increase of solute concentration during the freezing process, by almost an order of magnitude, compared to the isobaric cryopreservation. Isochoric and hyperbaric methods follow the same theoretical principles (changes in pressure), but they are different processes. The isochoric cryopreservation is a two-phase equilibrium process, in which ice and liquid exist simultaneously at equilibrium under constant temperature and volume, while hyperbaric cryopreservation the solution is maintained in a single phase, as liquid. In hyperbaric cryopreservation, the elevating pressure changes and rapid cooling rates prohibit the ice formation in the solution, thus the biological sample maintains its original composition. However, cells and organs cryopreserved by the hyperbaric method do not show very high survival rates. Isochoric cryopreservation offers a more robust method of freezing, by which the system adjusts itself to the minimal pressure for a particular temperature, instead of continuously increasing the pressure of the chamber. Erythrocytes in 5% v/v Me_2SO or 8% v/v glycerol solutions, under 120 MPa pressure and cooling rate of 35 °C/min or 160 °C/min showed survival rates of 92% or more.

Cryopreservation Equipment

LN_2: Direct Temperature Feedback Method

These systems monitor and control the temperature of the chamber. More advanced systems are additionally monitoring the samples' temperatures and they adjust the chamber's operational temperature accordingly. T-type thermocouples are continuously monitoring and controlling the chamber's and samples' temperatures, compare them to the programmed optimal temperatures and the appropriate amount of LN_2 is determined and injected into the chamber. The delivery of the LN_2 in the chamber is provided usually by dual injection valves for more precise temperature control, faster cooling rates and as a back-up system in case of failure of one of the valves. Solenoids can also be used, but they tend to wear over time.

LN_2: Timed Pulse Method

Similarly to direct temperature feedback systems, timed pulse systems control the chamber temperature. They usually consist of a microprocessor control system, a pump, a heater and one or more solenoid valves with openings, which meter a timed pulse of LN_2 into the chamber. The appropriate amount of LN_2 is determined and injected into the chamber by several variables, such as the valves' sizes, the tank's pressure and the number of solenoid openings. The heater heats the LN_2 to its boiling point, in order to provide constant tank pressure and to ensure that the appropriate amount of LN_2 is provided to the chamber. The disadvantage of timed pulse systems is their low reliability, as the same cooling program can produce different cooling curves of the cells over time.

LN_2: Plunge Freezing Method

This is the simplest method of controlled rate freezing with the use of LN_2 and it is usually used for small numbers and low volume straws and vials. A heat block is loaded with the biological samples and then is submerged into a tank of LN_2. The heater increases or decreases its power in order to achieve a controlled freezing rate of the chamber.

Step Down Method - Mechanical Freezers

This is a non-automated freezing method, but is still in practise. The samples are initially cooled down by being placed in a refrigerator for several hours, then transferred to a mechanical freezer and finally stored either into a LN_2 freezer or continue their storage into an Ultra-Low Temperature (ULT) freezer.

ULT freezers follow the principles of any mechanical refrigeration system and consist of the same basic components; a compressor, a preservation area, evaporator, condenser and refrigerant. The refrigeration of the system is provided by the basic vapour-compression cycle, in which the refrigerant circulates through the system and passes through a number of processes. The temperature range of an ULT freezer is typically between -56 to -86 °C, but it is important to be able to operate within the range of -70 to -80 °C. There is no conventional refrigerant that can cover the temperature gap between ambient room and operational temperatures; for that reason ULT freezers are equipped with two refrigeration circuits, each one with its individual compressor and refrigerants, using a cascade technique. Based on the cascade technique, the first refrigeration cycle is used to cool down the system from ambient to an intermediate temperature, while the second refrigeration circuit is able to further reduce the temperature of the preservation area to the required operational conditions.

The disadvantages of step down method are that is a long time process, is difficult to repeat and cannot provide controlled cooling rates. Moreover, the post-thaw recovery rates of cells maintained into ULT are very low, even for robust cells, due to power outages, mechanical failures or human caused mishaps, such as frequent door openings or improper closing of the freezer's door. As far as energy consumption is concerned, ULT systems account for more than 50% of the total energy used in a laboratory space.

Alcohol Bath Freezers

These units are commonly used in laboratories, for preservation of cells at temperatures above -80 °C. Alcohol bath freezers are basically refrigerated circulators, in which an amount of alcohol is used as cooling medium to pass through a cooling system and exchange temperature, in order to reduce it. These systems require a relatively large volume of cooling medium, in order to provide stable temperatures and accurate controlled cooling rates, due to medium's high heat capacity. However, alcohol bath freezers are not capable of achieving rapid cooling rates.

Vitrification Systems

These systems usually include the rapid transition of the sample to a glassy state, by direct submersion into LN_2. The cooling rate of the equipment can be increased either by the use of negative pressure systems, which depress the freezing point of LN_2 to -205 °C, or by minimising the sample's volume. Other available systems freeze the cell by getting them into contact with an aluminium block cooled in LN_2, in order to avoid the direct contact of the sample and the LN_2, for contamination reasons and to eliminate the Leidenfrost effects.

Stirling Engines Freezers

The original idea of Stirling engines is to convert thermal energy to mechanical work. However, their operation principle can be used in reverse and thus cryopreservation systems use Stirling engines to convert mechanical energy to thermal energy; in other words, Stirling engines operate as heat pumps. Most refrigeration systems operate on the Rankine cycle and their cooling capacity depends on the properties of their refrigerant. Triple stage Rankine refrigerators are currently the state of the art of mechanical freezers, but they can achieve temperatures of only -140 °C. Sterling cycle is by definition less efficient than Rankine cycle, but is capable of providing lower temperatures than the Rankine cycle. Moreover, they can freeze small amounts of samples and their cooling capacity is much lower than the LN_2 systems. However, due to personal safety and cross-contamination concerns of LN_2 systems, Sterling engines freezers find great application in clean rooms and laboratories where the use of LN_2 is not possible.

Liquidus Tracking Freezers

LT method is relatively a novel technique, thus there are not many types of equipment available. The typical configuration is a conventional slow cooling freezer coupled with a LT controller and two dynamic peristaltic pumps. The cooling is provided by the same way as in the conventional slow rate freezing process, but the LT controller monitors the temperature inside the chamber and adjusts the speed of the peristaltic pumps, in order to alter the concentration of CPA solution of the cells. The decrease of the temperature in the chamber, increases the CPA concentration of the cell, but as the temperature of the cell decreases, the toxicity of the CPA decreases, allowing the sample to tolerate the high CPA concentrations. The only requirements of the LT freezers are the large volumes of CPAs and their proper mixing around the sample.

Isochoric Systems

Isochoric systems are quite simple and inexpensive, however they are not commercially available. One of the systems described in the literature was used by Rubinsky et al. and it consists of a constant volume chamber that is hermetically sealed and in which the pressure is monitored with a pressure gauge. The chamber is filled with fluid and is cooled by immersion in a constant temperature bath. The major advantage of isochoric cryopreservation is that the solution concentrations during the freezing are lower at each temperature by almost an order of magnitude, compared to the isobaric cryopreservation; a fact that may increase the post recovery success rates of the cells.

Long-Term Storage in LN_2: Contamination and Safety Issues

The temperatures at which biological materials are stored have great influences on their shelf life. Generally, the lower the storage temperature, the longer the length of time after which the biological material can be recovered. Ultimate cell viability and stability is achieved, when samples are preserved below -130 °C the glass transition temperature of water. It is believed that at this temperature all metabolic processes are retarded and biological stability is obtained, which makes possible the intact storage of samples even for 2,000 years. The only consideration is the slow accumulation of background ionizing radiation, but the effects on the cells will appear after centuries of storage.

Storage at temperatures of -150 °C is typically recommended, because it provides a 20 °C safety margin between the storage and glass transition temperature of cells. Storage in LN_2, with a boiling point of -196 °C, is a convenient method to maintain low operating temperatures and to provide even greater safety margin for better sample security. The reasons of why LN_2 finds such great application in cryopreservation systems are due to the extremely low cryogenic temperatures that offers, the fact that it is chemically inert, relatively inexpensive, readily available, non-flammable and does not require electrical power to provide freezing; power supply is only used for the electronic features of the systems, such as auto-filling, monitoring and alarms.

However, besides the advantages of LN_2 systems, it is important to outline that the exposure to an environment with high levels of LN_2 in the atmosphere can cause dizziness, unconsciousness and even death, as the molecules of nitrogen displaces the molecules of oxygen. Thus the handling of LN_2 requires extra care, as it is colourless, odourless, and tasteless and it is breathed as if it were air, making its detection by the human senses impossible.

LN_2 storage systems can provide cooling in either liquid or vapour phase. Storage in liquid LN_2 offers uniform temperature profile across the chamber, while the vapour phase storage shows a temperature gradient and relatively higher temperatures than the liquid phase. However, storage in liquid phase LN_2 is contains the risk of contamination and explosion of the samples, as the vials used for packing the biological materials may leak and allow the interaction between the LN_2 and material. Thus, the rest of the samples stored in the chamber can become cross contaminated with microbes and LN_2 can enter the leaking vial. As a result, when the leaking sample is retrieved, the trapped LN_2 will expand, transform into gas, change its volume and cause a small explosion, and the destruction of the sample.

LN_2 usually has a very low microbial count but, during storage and distribution may become contaminated with microorganisms and potentially infect the stored material by transmitting viruses, bacteria, fungi and animal cells in the holding tank. By opening the liquid nitrogen container, the water vapour in the air above the container becomes cooled and forms small ice crystals with high electrostatic charge, which capture airborne microorganisms and fall into the container. However, the contamination can be controlled by the appropriate emptying and cleaning of the vessel. Other potential contaminations can occur from contaminated surfaces or leaking samples.

LN_2 should be treated as biohazardous substance, as the screw cap of the plastic vials allowed contact between the contaminated LN_2 and the sample. Another example of contamination is the Hepatitis B Virus (HBV) outbreak caused by LN_2 storage. In 1995, Tedder et al. reported how the bone marrow and stem cells harvested from one patient with HBV led to infect other five patients, who received transplants stored in the same tank. Also, a survey of microbial contamination of hematopoietic stem cells stored are LN_2 freezers was conducted by Fountain et al. and showed that of the 583 cultures tested, 1.2% were found to be contaminated by microorganisms. Four out of five freezers examined contained low level of microbial contamination, however the fifth freezer was heavily contaminated with Aspergillus, the same microbes found in the contaminated samples.

A more recent assay of Grout and Morris showed that small particles, can be transmitted in the vapour used to cool programmable vapour LN_2 freezers, and as the immediate storage environment for bulk storage and transport of cryopreserved samples and remain suspended in the vapour phase of LN_2 for at least 24 hours. The longer a sample has been held in a storage tank then the greater the risk of contamination from the accumulating microbial load of the bulk cryogen.

Attempts to eliminate the danger of cross contamination include cooling in LN_2 passed through a 0.2μm pore-size filter and placing the carrier (vial, straw, etc.) into a container that partially or completely isolates it from the LN_2 during storage.

Storage in vapour LN_2 can address the contamination and the explosion problems, however cannot provide the uniform temperature distribution in the chamber, developing that way different temperature gradients along the storage area. However, new designs of vacuum insulated chambers, with low evaporation rates of LN_2 are showing promising results and vapour LN_2 temperatures of -190 °C. These vapour LN_2 systems restrict the amount of energy entering the preservation area, ensuring that the temperature of the vapour is maintained close to the liquid LN_2 temperature, and the heat exchange surface extends from the bottom to the top of the freezer.

CRYOPRESERVATION OF HEMATOPOIETIC STEM CELLS

The role of hematopoietic stem cell transplantation in the treatment of hematologic and nonhematologic malignancies is rapidly expanding. In certain situations fresh stem cells can be employed in the setting of allogeneic transplantation. If the transfer from donor to recipient can be established within 72 hr, protocols for preliminary storage at suprafreezing temperatures are in place. However, the current therapeutic strategies demand that the progenitor cells are cryopreserved for virtually all autologous and many allogeneic transplants. This strategy has been proven to be safe and not associated with significant adverse outcomes regarding failure to engraft, graft versus host disease (GVHD), or engraftment failure.

The cryopreservation process is of importance for all types of stem cell collection, but is perhaps particularly critical for umbilical cord blood (UCB). The actual transplant is here harvested at the time of birth and used at a later point in time for a yet, at the time point of the harvest, often indeterminate recipient. The UCB is usually stored by either public or private cord blood banks. Public cord blood banks are usually nonprofit organizations, which offer the donor unit to matching recipients via national or international registries to potential recipients in need. Cord blood banks store a donor specimen for the donor or in the case of public banks for an unknown recipient for an indeterminate time span. There are now about 170,000 frozen units in 37 cord-blood registries in 21 countries. Two thousand nine hundred units have been transplanted to date, with adults having received about one third of those units.

The cryopreservation process entails the following general components:

- Harvesting of the donor cells, which entails the actual collection of the specimen and the reduction of bulk,
- Addition of cryopreservatives,
- The actual freezing procedure,
- Assessment of the viability of the frozen unit after about 72 hr,
- The thawing procedure,
- The washing and conditioning of the donor unit prior to transplant.

No single cryopreservation method has been universally used. Variations in technique occur between different transplant centers.

Temperature

The temperatures used for the cryopresevation of hemopoeitic stem cells over the past fifteen years has been −196, −156, or −80 °C, reflecting the storage temperatures in liquid and vapor phase nitrogen and in cryopreservation mechanical freezers, respectively. The development of the use of cryopreservative temperatures was from lower temperatures of around −196 °C in the 1980s to around −80 °C in the 1990s. For umbilical cord isolated stem cells similar trends are observed. The standard temperatures currently in use are −196 to −80 °C. Secondary to the recent reports about the spread of infectious agents (i.e., aspergillus as well as viral spread), through the liquid phase of the nitrogen tanks, the currently recommended optimal storage conditions are in the vapor nitrogen phase, at −156 °C. Mechanical freezers might represent a viable alternative.

Also, several studies examined the possibility to store HSCs at suprafreezing temperature, at 4 °C. A preclinical study that examined PBSCs mobilized in autologous plasma with post-storage clonogenic and viability assays suggested that a storage up to five days is safe. A small case series by Ruiz-Arguelles et al. successfully used PBPCs after 96 hr storage at 4 °C for rescue after high dose chemotherapy.

Freezing Rate

The rate of freezing was widely debated in the literature. The controlled rate freezing technique is still considered standard, mainly due to the fact that the heat liberation at the transition or eutactic point (about 4 °C) is deemed detrimental to the stem cell population. At this point the water molecules within the frozen unit are in a precise molecular order, what eventuates in the thermodynamic liberation of fusion heat.

In controlled rate freezing, the concentrated stem cells are frozen down at a rate of 1–2 °C/min up to a temperature point of about −40 °C. Then, the freezing process down to a target of −120 °C is performed at a faster pace, about 3–5 °C/min. For umbilical cord stem cells, bone marrow, and PBSCs the controlled rate freezing process is considered standard, and was in different reports found to be superior to uncontrolled freezing approaches. This procedure is time consuming and requires staff with a specific expertise. Hence, the use of uncontrolled rate freezing in which the specimen is first cooled down to −4 °C and then directly deposited into a freezer at −80 °C or put into liquid phase nitrogen has been evaluated. Several reports established that the uncontrolled method is safe and reveals comparable results to the controlled rate process for BM and PBSCs. A controlled study performed by Perez-Oteyza et al. showed that the controlled and uncontrolled rate freezing approach are comparable in terms of viability testing and that only a statistically significant decrease in the CFU-GM clonality assay could be detected in the uncontrolled freezing situation. Recent studies suggested that uncontrolled freezing is also a viable approach for UCB stem cells. No consensus exists about the relevance of the compensation for fusion heat during the freezing procedure. However, a study of Balint et al. outlined the importance of this intervention, comparing five different freezing protocols. The protocols using a five step controlled freezing approach compensating for the fusion heat achieved better post thawing viability.

Durability

The actual durability, defined as the time that stem cells can be preserved, is still unclear. The viability of the stem cells in cryopreservation has been questioned in different studies after a time course of cryopreservation of 6 months. Further studies have demonstrated a prolonged freezing time with complete preservation of stem cell function.

Several substitute assays are used to estimate the functional hematopoietic reconstitutive capacity of the first frozen and then thawed specimens. While BFU-E and CFU-GM appear to be compromised earlier in the course of cryopreservation, the recovery of nucleated cells (NC) and CD34+ cells and the actual engraftment in NOD/SCID mice seems to be preserved for a longer period of time. Those observations have been initially made in bone marrow and PBSC, and similar observations have been made with UBC stem cells. The NOD/SCID mouse assay is currently considered the most valuable assay to assess the functionality for hematologic recovery of HSC preparations, but is not routinely practical. After Kobylka et al. and Mugishima et al. proved the durability after 12 and 15 years with flow cytometric and clonogenic assays, respectively, and Broxmeyer et al. performed a reassessment of his long-term preserved CB units, a durability of up to 15 years was established by using hematopoietic reconstitution in NOD/SCID mice. Clinical validity of preclinical studies was documented in anecdotal reports when successful trilineage engraftment was achieved with BM, stored for 7 years. A systematic review evaluating the combined experience of the Brigham and Women's Hospital and the EBMT Group noticed that HSC can be effectively cryopreserved for up to 11 years. A retrospective study from Seattle revealed full trilineage recovery in patients receiving HSC, stored for up to 7.8 years without consistent detrimental effects.

Cell Concentration

Reinfusion of cryopreserved cells has been associated with varying toxicities, which were partially attributed to the total volume and the cryopreservatives in the solution. Concern was raised in the past that a high cell concentration in the cryopreservate can eventuate in toxicity to the cells. Hence, the initially proposed concentration of cryopreserved cells was suggested to be not over 2×10^{-7}/ml NC. This would eventuate in a cryopreservation volume of about 7 l per patient. The storage space needed and the labor to wash the graft prior to reinfusion would be immense.

After initial murine models were found safe, Rowley et al. established that high cell concentrations (up to 5.6×10^{-8} cells/ml) in the cryopreservate are well tolerated, not associated with significant adverse effect to the cells, and resulted in good clinical outcomes. Similar conclusions were drawn from subsequent studies by Kawano et al. and Cabezudo et al. For practical purposes a cell concentration of 200×10^{-6}/ml appears achievable.

Cryopreservatives

Cryopreservatives are necessary additives to stem cell concentrates, since they inhibit the formation of intra and extracellular crystals and hence cell death. The standard cryoprotectant is DMSO, which prevents freezing damage to living cells. It was initially introduced into medical use as an anti-inflammatory reagent and is still occasionally used in auto-immune disorders. Usually it is used at concentrations of 10% combined with normal saline and serum albumin. This was established to be a safe and non-stem cell toxic agent. However, DMSO is associated with a clinically significant side effect profile. Nausea, vomiting, and abdominal cramps occur in about half of all the cases. Other side effects encompass cardiovascular, respiratory, CNS, renal, hemolytic, and hepatotoxic presentations. Case fatalities attributed to DMSO toxicities have been reported.

A recent multinational questionnaire based survey, including data from 97 EBMT transplant centers, revealed that DMSO related toxicities other than nausea and vomiting are observed in about one out of 50 transplants with a mean incidence of 2.2% of all administered units. Cardiovascular side effects were the most frequently observed group of adverse events witnessed in 27% of the participating centers. Respiratory events were observed in 17%, CNS toxicities, including seizures, in 5%, and renal adverse effects in 5%.

Based on these toxicity considerations, newer approaches have been tried. Lower dosages of DMSO, varying from 2.2 to 6% have been established to be efficacious in bone marrow and PBSCs as well as for UCB. On the contrary, a Yugoslavian study compared the 10% DMSO concentration to lower concentrations with different freezing rates. The 10% DMSO cryopreservation proved to be superior to lower concentrations in this in vitro study. To enhance the effect of the cryopreservative, the combination of DMSO and the extracellular cryoprotectant hydroxyethyl starch (HES) has been used with success in PBSCs and bone marrow grafts and UCB cells.

Alternative preservation methods for cryopreservation are propylene glycol, a combination of alpha tocopherol, catalase, and ascorbic acid and the glucose dimer trehalose as intra and extracellular cryoprotectant.

Interesting preclinical data from Germany suggests that activation of caspases, particularly during the thawing process, can induce apoptosis and hence contribute to the cryoinjury to transplant grafts. Addition of the caspase inhibitor zVAD-fmk as cryopreservative presents an intriguing future perspective.

Thawing

Several techniques for the thawing procedure have been proposed. The standard method is warming in a water bath at 37 °C until all ice crystals disappear. A German study compared the thawing of cryopreserved units in a warm water bath with dry heat applied by gel pads at 37 °C. The viability and clonogenic potential were comparable, with a trend towards less infectious contamination in the dry method. Different studies examined the preservation of function when thawed units were incubated at 0–37 °C. No significant differences were detected in a study by Yang et al. who compared an incubation of the thawed unit at 0, 20, and 37 °C for 20 min. The used cryopreservative proved to be nontoxic to the stem cells during the cryopreservation process, as already established by previous studies. Reducing the DMSO content at thawing temperature is an intriguing concept, because of the clinical toxicity profile of this cryopreservative. Hence, the effect of reducing the DMSO content in the thawing solution by virtue of washing or dilution has been explored. Minor or no effect on the stem cell viability has been observed. An automated method to wash out the cryopreservative has proven feasible in pre-clinical models.

Washing Procedure

For stem cells of cord, peripheral blood, and bone marrow origin, the process of washing out the cryopreservative after the thawing can still be considered standard, since the DMSO is assumed to have toxic effect on the stem cells. This has been questioned by several more recent reports, which suggested stem cell resistance against DMSO exposure. The wash out of the most popular cryopreservative has conceivable benefits for the recipient, i.e. reduction of toxicity, since the degree of DMSO toxicity is proportional to the amount of DMSO contained in the infused stem cell solution. It was also suggested that wash out of DMSO can enhance engraftment. This has been disputed.

The current standard washing protocol follows the New York Blood Center protocol, in which the two step dilution of the thawed stem cell unit with 2.5% human serum albumin and 5% dextran 40 is followed by centrifugation at 10 °C for 10 min. The supernatant is then removed and HSA and dextran solution is again added twice to a final DMSO concentration of less then 1.7%. The washed solution is infused as soon as possible. Although this procedure has been established to be safe and associated with a reasonable recovery of NC and progenitor assays, it is also very labor intensive and not free of cell loss. Recently new automated cell washing devices have been introduced with promising results.

Containers for Storage

The International Society for Cellular Therapy (ISCT) described on its supplier information website for cryocontainers nine different cryostorage container products. Six of them are Ethinyl Vinyl Acetate (EVA) based, usually gamma irradiated.

Trademarks are:

- Cryocyte/Baxter,
- CellFlex/Maco Pharma,
- Cell Freeze™ Charter Medical,
- Pall Medical Freezing Bag 791-05,
- Cryostore EVA/Origen Biomedical Inc.,
- Thermogenesis Corp./Freezing bag 80346-0.

Other, not EVA based products are:

- American Fluoroseal/FEP(Teflon),
- Fresenius Hemocare/Hemofreeze(Teflon,Kaplon),
- Origen Biomedical Inc./Permalife Bag, FEP/Polyimide.

Other approaches have been undertaken. A Czech group published their successful experience with a stainless steel cryopreservation container specifically designed for PBSC. In the US, the most commonly used cryocontainer is an EVA freezing bag.

The use of specific containers, PVC and polyolefin plastic bags and polyethelene cryostorage vials, achieved different results regarding the viability of the stored specimen. PVC and polyolefin bags achieved satisfactory results, while polyethelene cryostorage vials did not in one study. A group from Boston suggested that polyolefin cryobags achieve a longer functional duration than PVC bags.

Infectious Considerations

Microbial contamination of transplants represents a significant hazard to the severely immunosuppressed recipient. The FDA estimates that seven transplant related deaths per year could be avoided by elimination of infection related to donor cell infusions. The overall demonstrated microbiological contamination rates are 0–4.5%.

The major parts of the cultured bacteria are skin flora and commensal bacteria. The remainder is mainly made up out of enteric bacteria. Of note is that bone marrow derived stem cell cryounits are much more likely to be contaminated by pathogens, which is explained by the harvesting process. The rates of contamination between PBSC and bone marrow differ significantly by up to a factor of sixteen (0.23% for PBSC and 3.8% for BM).

The table below displays the incidence with which different pathogens were cultured from donor units in four different studies addressing the bacterial contamination of stem cell products:

Table: Organisms cultured/overall incidence of positive cultures.

Organisms cultured	Overall incidence of positive cultures (%)
Staph. epidermidis and other coagulase negative Staphylococcus (CNS)	3–11.7
Propionibacterium acni	0.6–2.2
Staphylococcus aureus	0–1.6
Bacillus cereus and other Bacillus spec.	0.06–0.35
Pseudomonas spec.(aeruginosa, putida and fluoresces)	0.1–0.8
Corynebacterium spec.	0–0.3
Aspergillus fumigatus	0–0.3
Mixed cultures	0.1–1.6

The cryopreservation process was associated with reduction of detectable microorganisms. In one German study the detection of Staphylococcus epidermidis was reduced by an average of 9.3% and Escherichia coliby 18.1%. Also, several studies reported the occurrence of positive cultures post thawing. This suggests the risk of contamination of the culture bottles (i.e., not induced by the donor graft).

The incidence of severe sepsis upon infusion of stem cells, which cultured positive for commensals and skin flora bacteria, is low and most of the febrile episodes developing after their infusion are treatable with antibiotics.

The stability of viruses in liquid nitrogen has been documented. An English source published an epidemic outbreak of hepatitis B in autologous bone marrow recipients, which was subsequently linked by nucleotide sequence analysis to another cryopreservative stored in the same container. Subsequent analysis of debris in the liquid nitrogen phase of the same container demonstrated spread of the pathogen via the liquid phase. Similar outbreaks have been reported. To prevent such incidences we store infectious conserves separately and provide protective sleeves around the cryopreservative bags, as reported to be efficacious previously.

To prevent infectious complications by the infusion of donor stem cells the following measures should be employed:

- Processing in clean areas and scrupulous microbiologic monitoring of all stages of the stem cell preservation procedure according to current standards.
- Detection of microbiologic contamination prior to infusion. This has been successfully done in an automated manner for other blood products.
- Screening of donors, even in the autologous setting as circulated in regular guidelines. Upon detection of an infectious graft there should be separate or protected storage.

Embryonic Stem Cells

Embryonic stem cells portray a different biologic behavior under cryopreservation. Because of the enormous potential of Embryonic Stem (ES) cells for transplantation therapy, recent studies have evaluated the manner in which these fragile cells are stored. One difficulty with the cryopreservation of these cells is their extreme fragility resulting in poor survival of the cells under standard freezing procedures, usually in the range of 1%. Not only do the cells have a poor yield with the standard freezing protocols, the cells are also induced into differentiation. Ware et al. investigated a method using a very slow controlled rate freezing procedure and 10% DMSO. Along with the very slow freezing rate, a rapid thaw was found to be critical for a successful storage. A study from Wisconsin identified HES (hydroxy ethyl starch) as a valuable cryoadditive during slow rate freezing and vitrification procedures. Some methods have been derived in which the ES cells are frozen in 24-well plates with minimal media and 10% DMSO at −80 °C. The importance of rapid thawing by adding minimal media at body temperature was emphasized. In a study by Ure et al., all 227 clones tested grew successfully, although molecular and phenotypic studies were not done in this instance to prove the cell lineage. Adams et al. successfully cryopreserved primary hepatocytes in a University of Wisconsin solution containing FBS and DMSO for 8 months with the preservation of viability and key phenotypic properties. A recent preclinical study by Milosevic et al. highlighted the role for caspase inhibitors as additives to cryopreservative in embryonic murine neural precursors, an approach that was previously undertaken in hematopoietic precursors. A 60–70% viability was achieved by adding the caspase inhibitor zVAD-fmk to different cryopreservatives after five days.

One theory as to the cause of cell death is ice crystal formation in the cytoplasm during freezing. The process of vitrification makes attempts of freezing the ES cells while avoiding the ice formation. Reubinoff was the first to implement this method using an open pulled straw vitrification method, which had previously been successful in the cryopreservation of embryos. This procedure, which evaluated the more fragile human ES cells, proved a 100% viability of the ES cell clumps that all generated colonies compared to a 70% recovery post thaw and 16% differentiation using standard methods. The test cells also had normal karyotypes, OCT-4 expression, and developed teratomas in xenografts of SCID mice. Since this experiment, others have looked at closed seal straw vitrification or alternative freezing media and a simplified vitrification method that showed similar results.

Functional Substitute Assays

The most commonly used clonogenic assays are CFU-Sd12, a murine assay, MRA (CFU-GM), CFU-GEMM, BFU-E, and the long term culture initiating cell (LTC-IC) assay. Still, in spite of the availability of these assays to quantitatively and qualitatively assess the clonogenic potential of the hemopoietic cell in suspension, the eventual evaluation of the engraftment potential relies on the evidence of hematopoietic reconstitution in myeloablated mammals.

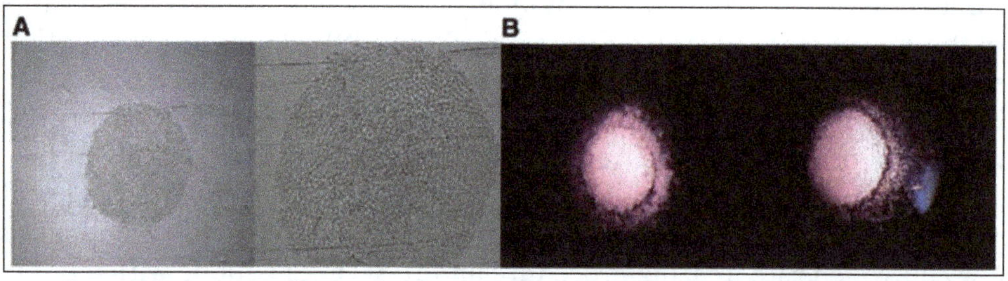

Clonogenic assays. Cells are plated in soft agar and incubated. At set time periods, the cell colonies are counted; this provides an in vitro surrogate of hematopoietic reconstitution potential. (A) CFU/GM (B) CFU/GEMMA.

Different techniques that have been used for human cells are cell counting for NC and CD34+ cells, tryptan blue exclusion for viability, 7-Aminoactinomycin, engraftment in NOD/SCID mice, and clonogenic assays.

Though no absolute consensus is reached as to the optimal method to assess the functionality of a donor graft, several substitute assays have been used to estimate the functional hematopoietic reconstitutive capacity of the first frozen and then thawed specimen. The broad categories of assays used for this purpose are cell counting assays, viability assays, clonogenic experiments, and the engraftment of donor cells in NOD/SCID mice.

1. Cell counting assays:
    - Count of NC,
    - Flow cytometry of CD34+ cells.
2. Viability assays:
    - Tryptan blue,
    - 7-Aminoactinomycin,

- Propidium iodide.

3. Clonogenic assays:
    - CFU-Sd12 assay in mice,
    - CFU-GM,
    - CFU-GEMM,
    - BFU-E,
    - LTC-IC.
4. Direct engraftment experiments:
    - NOD/SCID mice.

While the BFU-E and CFU-GM appear to be compromised earlier in the course of cryopreservation, the recovery of NC and CD 34+ cells and the actual engraftment in NOD/SCID mice seems to be preserved for a longer period of time. These observations have been made in bone marrow and PBSC, and for UBC the same observations were made.

Yang et al. evaluated two different functional assays, the CD 34 count and the CFU-GM, by correlating the pre and post thawing assay outcome with engraftment in 52 patients. The pre and post thawing assay correlated well with each other and with the actual clinical engraftment. A general summary of a standard cryopreservation technique is presented in the figure below.

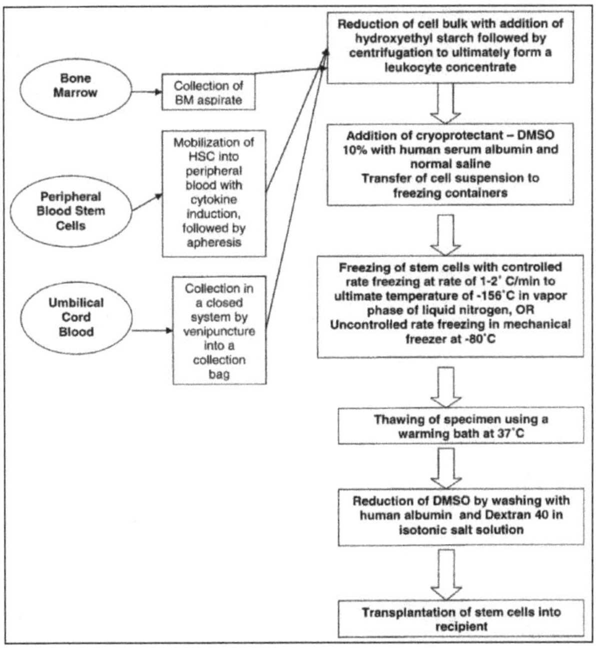

The cryopreservation process for bone marrow, peripheral blood stem cells, and UCB.

References

- Cryopreservation, technology: britannica.com, Retrieved 27 May, 2019
- Precision-in-cryopreservation-equipment-and-control, current-frontiers-in-cryobiology: intechopen.com, Retrieved 05 July, 2019
- Di Santo, M; Tarozzi, N; Nadalini, M; Borini, A (2012). "Human Sperm Cryopreservation: Update on Techniques, Effect on DNA Integrity, and Implications for ART". Advances in Urology. 2012: 854837. doi:10.1155/2012/854837. PMC 3238352. PMID 22194740
- Edgar, D. H.; Gook, D. A. (2012). "A critical appraisal of cryopreservation (slow cooling versus vitrification) of human oocytes and embryos". Human Reproduction Update. 18 (5): 536–554. doi:10.1093/humupd/dms016. PMID 22537859
- Cryopreservation-of-testicular-tissue, current-frontiers-in-cryobiology: intechopen.com, Retrieved 17 July, 2019
- Isachenko V, Lapidus I, Isachenko E, et al. (2009). "Human ovarian tissue vitrification versus conventional freezing: morphological, endocrinological, and molecular biological evaluation". Reproduction. 138 (2): 319–27. doi:10.1530/REP-09-0039. PMID 19439559
- Kopeika, J.; Thornhill, A.; Khalaf, Y. (2014). "The effect of cryopreservation on the genome of gametes and embryos: principles of cryobiology and critical appraisal of the evidence". Human Reproduction Update. 21 (2): 209–227. doi:10.1093/humupd/dmu063. ISSN 1355-4786. PMID 25519143

Cryogenics

4 CHAPTER

Cryogenics refers to the study of materials and their behavioral aspects due to production and maintenance at extremely low temperature. Some of the elements of cryogenics are cryogenic fuel, cryogenic energy storage, cryocooler, cryogenic processor, cooling bath, etc. The topics elaborated in this chapter will help in gaining a better perspective about cryogenics.

Cryogenics is the science of producing and studying low-temperature conditions. It has come to mean the generation of temperatures well below those of normal human experience. More specifically, a low-temperature environment is termed a cryogenic environment when the temperature range is below the point at which permanent gases begin to liquefy. Permanent gases are elements that normally exist in the gaseous state and were once believed impossible to liquefy. Among others, they include oxygen, nitrogen, hydrogen, and helium.

The origin of cryogenics as a scientific discipline coincided with the discovery by nineteenth-century scientists that the permanent gases can be liquefied at exceedingly low temperatures. Consequently, the term "cryogenic" applies to temperatures from approximately −100 °C (−148 °F) down to absolute zero (the coldest point a material could reach).

The temperature of any material—solid, liquid, or gas—is a measure of the energy it contains. That energy is due to various forms of motion among the atoms or molecules of which the material is made. A gas that consists of very rapidly moving molecules, for example, has a higher temperature than one with molecules that are moving more slowly.

In 1848, English physicist William Thomson (later known as Lord Kelvin; 1824–1907) pointed out the possibility of having a material in which particles had ceased all forms of motion. The absence of all forms of motion would result in a complete absence of heat and temperature. Thomson defined that condition as absolute zero.

Methods of Producing Cryogenic Temperatures

Cryogenic conditions are produced by one of four basic techniques: Heat conduction, evaporative cooling, cooling by rapid expansion (the Joule-Thomson effect), and adiabatic demagnetization. The first two are well known in terms of everyday experience.

The third is less well known but is commonly used in ordinary refrigeration and air conditioning units, as well as in cryogenic applications. The fourth process is used primarily in cryogenic applications and provides a means of approaching absolute zero.

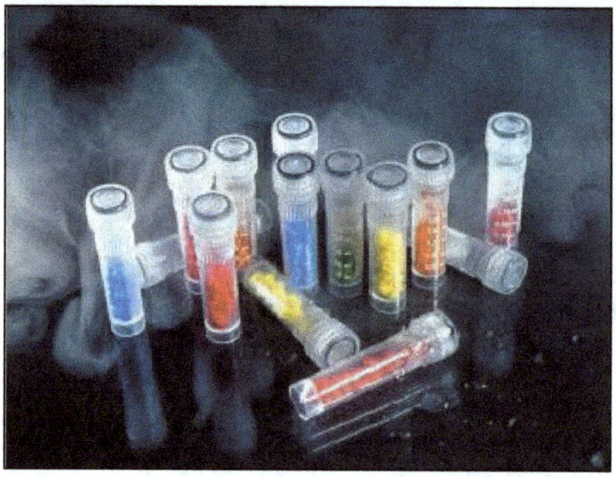

Cryotubes used to store strains of bacteria at low temperature. Bacteria are placed in little holes in the beads inside the tubes and then stored in liquid nitrogen.

Heat conduction is a relatively simple concept to understand. When two bodies are in contact, heat flows from the body with the higher temperature to the body with a lower temperature. Conduction can occur between any and all forms of matter, whether gas, liquid, or solid. It is essential in the production of cryogenic temperatures and environments. For example, samples may be cooled to cryogenic temperatures by immersing them directly in a cryogenic liquid or by placing them in an atmosphere cooled by cryogenic refrigeration. In either case, the sample cools by conduction (or transfer) of heat to its colder surroundings.

The second process for producing cryogenic conditions is evaporative cooling. Humans are familiar with this process because it is a mechanism by which our bodies lose heat. Atoms and molecules in the gaseous state are moving faster than atoms and molecules in the liquid state. Add heat energy to the particles in a liquid and they will become gaseous. Liquid perspiration on human skin behaves in this way. Perspiration absorbs body heat, becomes a gas, and evaporates from the skin. As a result of that heat loss, the body cools down.

Cryogens and their Boiling Points

Cryogen	Boiling Point		
	°F	°C	K
Oxygen	-297	-183	90
Nitrogen	-320	-196	77
Hydrogen	-423	-253	20
Helium	-452	-269	4.2

Neon	-411	-246	27
Argon	-302	-186	87
Krypton	-242	-153	120
Xenon	-161	-107	166

In cryogenics, a container of liquid is allowed to evaporate. Heat from within the liquid is used to convert particles at the surface of the liquid to gas. The gas is then pumped away. More heat from the liquid converts another surface layer of particles to the gaseous state, which is also pumped away. The longer this process continues, the more heat is removed from the liquid and the lower its temperature drops. Once some given temperature is reached, pumping continues at a reduced level in order to maintain the lower temperature. This method can be used to reduced the temperature of any liquid. For example, it can be used to reduce the temperature of liquid nitrogen to its freezing point or to lower the temperature of liquid helium to approximately 1 K.

A third process makes use of the Joule-Thomson effect, which was discovered by English physicist James Prescott Joule and William Thomson, Lord Kelvin, in 1852. The Joule-Thomson effect depends on the relationship of volume (bulk or mass), pressure, and temperature in a gas. Change any one of these three variables, and at least one of the other two (or both) will also change. Joule and Thomson found, for example, that allowing a gas to expand very rapidly causes its temperature to drop dramatically. Reducing the pressure on a gas accomplishes the same effect.

To cool a gas using the Joule-Thomson effect, the gas is first pumped into a container under high pressure. The container is fitted with a valve with a very small opening. When the valve is opened, the gas escapes from the container and expands quickly. At the same time, its temperature drops. The first great success for the Joule-Thomson effect in cryogenics was achieved by Kamerlingh Onnes in 1908 when he liquefied helium.

The Joule-Thomson effect is an important part of our lives today, even though we may not be aware of it. Ordinary household refrigerators and air conditioners operate on this principle. First, a gas is pressurized and cooled to an intermediate temperature by contact with a colder gas or liquid. Then, the gas is expanded, and its temperature drops still further. The heat needed to keep this cycle operating comes from the inside of the refrigerator or the interior of a room, producing the desired cooling effect.

The fourth process for producing cryogenic temperatures uses a phenomenon known as adiabatic demagnetization. Adiabatic demagnetization makes use of special substances known as paramagnetic salts. A paramagnetic salt consists of a very large collection of particles that act like tiny (atom-sized) magnets. Normally these magnetic particles are spread out in all possible directions. As a result, the salt itself is not magnetic. That condition changes when the salt is placed between the poles of a magnet. The magnetic field of the magnet causes all the tiny magnetic particles in the salt to line up in the same direction. The salt becomes magnetic, too.

At this exact moment, however, suppose that the external magnet is taken away and the paramagnetic salt is placed within a liquid. Almost immediately, the tiny magnetic particles within the salt return to their random, every-which-way condition. To make this change, however, the particles require an input of energy. In this example, the energy is taken from the liquid into which the salt was placed. As the liquid gives up energy to the paramagnetic salt, its temperature drops.

Adiabatic demagnetization has been used to produce some of the coldest temperatures ever observed—within a few thousandths of a degree kelvin of absolute zero. A related process involving the magnetization and demagnetization of atomic nuclei is known as nuclear demagnetization. With nuclear demagnetization, temperatures within a few millionths of a degree of absolute zero have been reached.

Applications

Following his successful liquefaction of helium in 1908, Kamerlingh Onnes turned his attention to the study of properties of other materials at very low temperatures. The first property he investigated was the electrical resistance of metals. Electrical resistance is the tendency of a substance to prevent the flow of an electrical current through it. Scientists had long known that electrical resistance tends to decrease with decreasing temperature. They assumed that resistance would completely disappear at absolute zero.

Research in this area had great practical importance. All electrical appliances (ovens, toasters, television sets, and radios, for example) operate with low efficiency because so much energy is wasted in overcoming electrical resistance. An appliance with no electrical resistance could operate at much less cost than existing appliances.

What Onnes discovered, however, was that for some metals, electrical resistance drops to zero very suddenly at temperatures above absolute zero. The effect is called superconductivity and has some very important applications in today's world. For example, superconductors are used to make magnets for particle accelerators (devices used, among other things, to study subatomic particles such as electrons and protons) and for magnetic resonance imaging (MRI) systems (a diagnostic tool used in many hospitals).

The discovery of superconductivity led other scientists to study a variety of material properties at cryogenic temperatures. Today, physicists, chemists, material scientists, and biologists study the properties of metals, as well as the properties of insulators, semiconductors, plastics, composites, and living tissue. Over the years, this research has resulted in the identification of a number of useful properties. One such property common to most materials that are subjected to extremely low temperatures is brittleness. The recycling industry takes advantage of this by immersing recyclables in liquid nitrogen, after which they are easily pulverized and separated for reprocessing.

Still another cryogenic material property that is sometimes useful is that of thermal contraction. Materials shrink when cooled. To a point (about the temperature of liquid nitrogen), the colder a material gets the more it shrinks. An example is the use of liquid nitrogen in the assembly of some automobile engines. In order to get extremely tight fits when installing valve seats, for instance, the seats are cooled to liquid nitrogen temperatures, whereupon they contract and are easily inserted in the engine head. When they warm up, a perfect fit results.

Cryogenic liquids are also used in the space program. For example, cryogenic materials are used to propel rockets into space. A tank of liquid hydrogen provides the fuel to be burned and a second tank of liquid oxygen is provided for combustion.

Another space application of cryogenics is the use of liquid helium to cool orbiting infrared telescopes. Infrared telescopes detect objects in space not from the light they give off but from the infrared radiation (heat) they emit. However, the operation of the telescope itself also gives off heat. What can be done to prevent the instrument from being blinded by its own heat to the infrared radiation from stars? The answer is to cool parts of the telescope with liquid helium. At the temperature of liquid helium (1.8 K) the telescope can easily pick up infrared radiation of the stars, whose temperature is about 3 K.

CRYOGENIC FUEL

Cryogenic fuels are fuels that require storage at extremely low temperatures in order to maintain them in a liquid state. These fuels are used in machinery that operates in space (e.g. rocket ships and satellites) because ordinary fuel cannot be used there, due to absence of an environment that supports combustion (on Earth, oxygen is abundant in the atmosphere, whereas in human-explorable space, oxygen is virtually non-existent) and space is a vacuum. Cryogenic fuels most often constitute liquefied gases such as liquid hydrogen.

Some rocket engines use regenerative cooling, the practice of circulating their cryogenic fuel around the nozzles before the fuel is pumped into the combustion chamber and ignited. This arrangement was first suggested by Eugen Sänger in the 1940s. The Saturn V rocket that sent the first manned missions to the Moon used this design element, which is still in use today.

Quite often, liquid oxygen is mistakenly called cryogenic *fuel*, though it is actually an oxidizer and not a fuel.

Russian aircraft manufacturer Tupolev developed a version of its popular Tu-154 design but with a cryogenic fuel system, designated the Tu-155. Using a fuel referred to as liquefied natural gas (LNG), its first flight was in 1989.

Operation

Cryogenic fuels can be placed into two categories: inert and flammable or combustible. Both types exploit the large liquid to gas volume ratio that occurs when liquid transitions to gas phase. The feasibility of cryogenic fuels is associated with what is known as a high mass flow rate. With regulation, the high-density energy of cryogenic fuels is utilized to produce thrust in rockets and controllable consumption of fuel.

Inert

These types of fuels typically use the regulation of gas production and flow to power pistons in an engine. The large increases in pressure are controlled and directed toward the engine's pistons. The pistons move due to the mechanical power transformed from the monitored production of gaseous fuel. A notable example can be seen in Peter Dearman's liquid air vehicle. Some common inert fuels include:

- Liquid nitrogen,
- Liquid air,
- Liquid helium,
- Liquid neon.

Combustible

These fuels utilize the beneficial liquid cryogenic properties along with the flammable nature of the substance as a source of power. These types of fuel are well known primarily for their use in rockets. Some common combustible fuels include:

- Liquid hydrogen,
- Liquid natural gas (LNG),

- Liquid methane.

Engine Combustion

Combustible cryogenic fuels offer much more utility than most inert fuels can. Liquefied natural gas, as with any fuel, will only combust when properly mixed with right amounts of air. As for LNG, the bulk majority of efficiency depends on the methane number, which is the gas equivalent of the octane number. This is determined based on the methane content of the liquefied fuel and any other dissolved gas, and varies as a result of experimental efficiencies. Maximizing efficiency in combustion engines will be a result of determining the proper fuel to air ratio and utilizing the addition other hydrocarbons for added optimal combustion.

Production Efficiency

Gas liquefying processes have been improving over the past decades with the advent of better machinery and control of system heat losses. Typical techniques take advantage of the temperature of the gas dramatically cooling as the controlled pressure of a gas is released. Enough pressurization and then subsequent depressurization can liquefy most gases, as exemplified by the Joule-Thomson effect.

Liquefied Natural Gas

While it is cost effective to liquefy natural gas for storage, transport, and use, roughly 10 to 15 percent of the gas gets consumed during the process. The optimal process contains four stages of propane refrigeration and two stages of ethylene refrigeration. There can be the addition of an additional refrigerant stage, but the additional costs of equipment are not economically justifiable. Efficiency can be tied to the pure component cascade processes which minimize the overall source to sink temperature difference associated with refrigerant condensing. The optimized process incorporates optimized heat recovery along with the use of pure refrigerants. All process designers of liquefaction plants using proven technologies face the same challenge: to efficiently cool and condense a mixture with a pure refrigerant. In the optimized Cascade process, the mixture to be cooled and condensed is the feed gas. In the propane mixed refrigerant processes, the two mixtures requiring cooling and condensing are the feed gas and the mixed refrigerant. The chief source of inefficiency lies in the heat exchange train during the liquefaction process.

Advantages and Disadvantages

Benefits

- Cryogenic fuels are environmentally cleaner than gasoline or fossil fuels. Among other things, the greenhouse gas rate could potentially be reduced by 11–20% using LNG as opposed to gasoline when transporting goods.

- Along with their eco-friendly nature, they have the potential to significantly decrease transportation costs of inland products because of their abundance compared to that of fossil fuels.
- Cryogenic fuels have a higher mass flow rate than fossil fuels and therefore produce more thrust and power when combusted for use in an engine. This means that engines will run farther on less fuel overall than modern gas engines.
- Cryogenic fuels are non-pollutants and therefore, if spilled, are no risk to the environment. There will be no need to clean up hazardous waste after a spill.

Potential Drawbacks

- Some cryogenic fuels, like LNG, are naturally combustible. Ignition of fuel spills could result in a large explosion. This is possible in the case of a car crash with an LNG engine.
- Cryogenic storage tanks must be able to withstand high pressure. High-pressure propellant tanks require thicker walls and stronger alloys which make the vehicle tanks heavier, thereby reducing performance and practicality.
- Despite non-toxic tendencies, cryogenic fuels are denser than air. As such, they can lead to asphyxiation. If leaked, the liquid will boil into a very dense, cold gas and if inhaled, could be fatal.

CRYOGENIC ENERGY STORAGE

Cryogenic energy storage (CES) is the use of low temperature (cryogenic) liquids such as liquid air or liquid nitrogen as energy storage. Both cryogens have been used to power cars. The inventor Peter Dearman initially developed a liquid air car, and then used the technology he developed for grid energy storage. The technology is being piloted at a UK power station.

Grid Energy Storage

Process

When it is cheaper (usually at night), electricity is used to cool air from the atmosphere to -195 °C using the Claude Cycle to the point where it liquefies. The liquid air, which takes up one-thousandth of the volume of the gas, can be kept for a long time in a large vacuum flask at atmospheric pressure. At times of high demand for electricity, the liquid air is pumped at high pressure into a heat exchanger, which acts as a boiler. Air from the atmosphere at ambient temperature, or hot water from an industrial heat source, is used to heat the liquid and turn it back into a gas. The massive increase in volume and pressure from this is used to drive a turbine to generate electricity.

Efficiency

In isolation the process is only 25% efficient, but this is greatly increased (to around 50%) when used with a low-grade cold store, such as a large gravel bed, to capture the cold generated by evaporating the cryogen. The cold is re-used during the next refrigeration cycle.

Efficiency is further increased when used in conjunction with a power plant or other source of low-grade heat that would otherwise be lost to the atmosphere. Highview Power Storage claims an AC to AC round-trip efficiency of 70%, by using an otherwise waste heat source at 115 °C. The IMechE (Institution of Mechanical Engineers) agrees that these estimates for a commercial-scale plant are realistic. However this number was not checked or confirmed by independent professional institutions.

Currently surplus gaseous nitrogen is produced as a byproduct in the production of oxygen. Oxygen can be used in oxy-combustion coal power plants, enabling CO 2 capture and sequestration. This gaseous nitrogen can be liquefied by available liquefaction capacities for further use. Cryogenic distillation of air is currently the only commercially viable technology for large scale oxygen production.

Pilot Plant

A 300 kW, 2.5 MWh storage capacity pilot cryogenic energy system developed by researchers at the University of Leeds and Highview Power Storage, that uses liquid air (with the CO_2 and water removed as they would turn solid at the storage temperature) as the energy store, and low-grade waste heat to boost the thermal re-expansion of the air, operated at a 80MW biomass power station in Slough, UK, from 2010 until 2014 when it was relocated to the university of Birmingham . The efficiency is less than 15% because of low efficiency hardware components used, but the engineers are targeting an efficiency of about 60 percent for the next generation of CES based on operation experiences of this system.

The system is based on proven technology, used safely in many industrial processes, and does not require any particularly rare elements or expensive components to manufacture. Dr Tim Fox, the head of Energy at the IMechE says "It uses standard industrial components - which reduces commercial risk; it will last for decades and it can be fixed with a spanner."

Grid-scale Demonstrator

Highview

In April 2014 the UK government announced it had given £8 million to Viridor Waste Management Ltd. and Highview Power Storage to fund the next stage of the demonstration. The resulting grid-scale demonstrator plant at Pilsworth Landfill facility in Bury, Greater Manchester, UK, started operation in April 2018.

This is based on research by the Birmingham Centre for Cryogenic Energy Storage (BCCES) associated with the University of Birmingham, and has storage for up to 15 MWh, and can generate a peak supply of 5 MW (so when fully charged lasts for 3 hours at maximum output) and is designed for an operational life of 40 years.

Tacoma Plant

In 2019 the Washington State Department of Commerce's Clean Energy Fund announced it would provide a grant to help Tacoma Power partner with Praxair to build a 15 MW/450 MWh liquid air energy storage plant. It will store up to 850,000 gallons of liquid nitrogen to help balance power loads.

CRYOCOOLER

A Cryocooler is a standalone cooler, usually of table-top size. It is used to cool some particular application to cryogenic temperatures.

Ideal Heat Exchangers and Regenerators

Heat exchangers are important components of all cryocoolers. Ideal heat exchangers have no flow resistance and the exit gas temperature is the same as the (fixed) body temperature T_X of the heat exchanger. Note that even a perfect heat exchanger will not affect the entrance temperature T_i of the gas. This leads to losses.

An important component of refrigerators, operating with oscillatory flows, is the regenerator. A regenerator consists of a matrix of a solid porous material, such as granular particles or metal sieves, through which gas flows back and forth. Periodically heat is stored and released by the material. The heat contact with the gas must be good and the flow resistance of the matrix must be low. These are conflicting requirements. The thermodynamic and hydrodynamic properties of regenerators are complicated, so one usually makes simplifying models. In its most extreme form an *ideal* regenerator has the following properties:

- Parge volumetric heat capacity of the material,
- Perfect heat contact between gas and matrix,
- Zero flow resistance of the matrix,
- Zero porosity (this is the volume fraction of the gas),
- Zero thermal conductivity in the flow direction,
- The gas is ideal.

Progress in the cryocooler field in recent decades is in large part due to development of new materials having high heat capacity below 10K.

Stirling Refrigerators

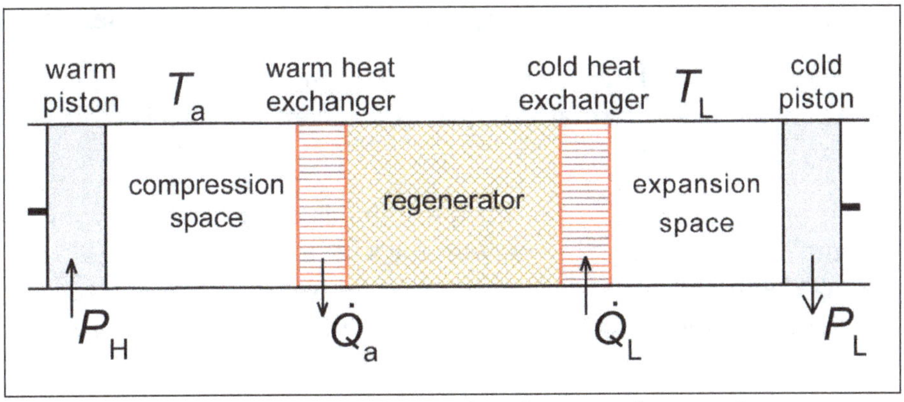

Schematic diagram of a Stirling cooler. The system has one piston at ambient temperature T_a and one piston at low temperature T_L.

The basic type of Stirling-type cooler is depicted in Fig. From left to right it consists of a piston, a compression space and heat exchanger (all at ambient temperature T_a), a regenerator, a heat exchanger, an expansion space and a piston (all at the low temperature T_L). Left and right the thermal contact with the surroundings at the temperatures T_a and T_L is supposed to be perfect so that the compression and expansion are isothermal. The work, performed during the expansion, is used to reduce the total input power. Usually helium is the working fluid.

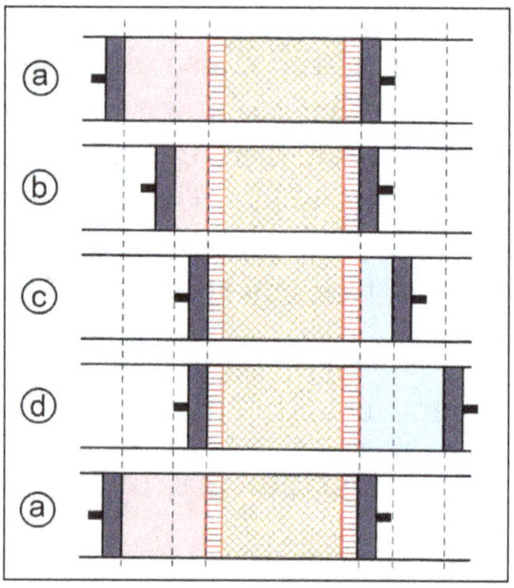

Four states in the Stirling cycle.

The cooling cycle is split in 4 steps as depicted in Fig. The cycle starts when the two pistons are in their most left positions:

- From a to b. The warm piston moves to the right while the cold piston is fixed. The compression at the hot end is isothermal (by definition), so heat Q_a is given off to the surroundings at ambient temperature T_a.
- From b to c. The two pistons move to the right. The volume between the two pistons is kept constant. The hot gas enters the regenerator with temperature T_a and leaves it with temperature T_L. The gas gives off heat to the regenerator material.
- From c to d. The cold piston moves to the right while the warm piston is fixed. The expansion is isothermal and heat Q_L is taken up. This is the useful cooling power.
- From d to a. The two pistons move to the left while the total volume remains constant. The gas enters the regenerator with low temperature T_L and leaves it with high temperature T_a so heat is taken up from the regenerator material. At the end of this step the state of the cooler is the same as in the beginning.

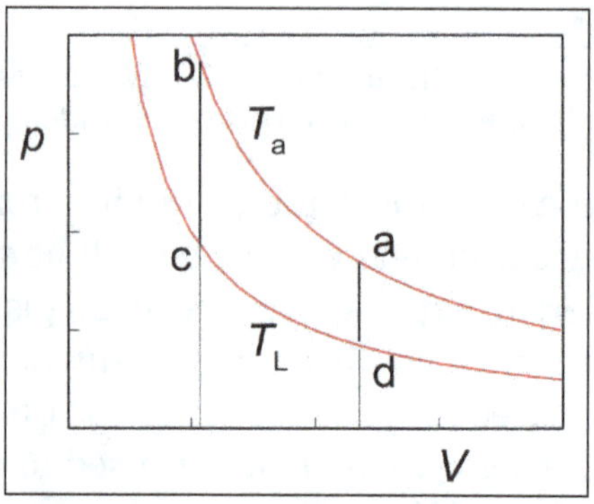

pV-diagram of the ideal Stirling cycle.

In the pV diagram the corresponding cycle consists of two isotherms and two isochores. The volume V is the volume between the two pistons. In practice the cycle is not divided in discrete steps as described above. Usually the motions of both pistons are driven by a common rotary axes which makes the motions harmonic. The phase difference between the motions of the two pistons is about 90°. In the ideal case the cycle is reversible so the COP (the ratio of the cooling power and the input power) is equal to the Carnot COP given by $T_L/(T_a - T_L)$.

It is not so practical to have a cold piston, as described above, so, in many cases, a displacer is used instead of the cold piston. A displacer is a solid body which moves back and forth in the cold head driving the gas back and forth between the warm and the cold end of the cold head via the regenerator. No work is required to move the displacer since, ideally there is no pressure drop over it. Typically its motion is 90 degrees out of phase with the piston. In the ideal case the COP also equals to the Carnot COP.

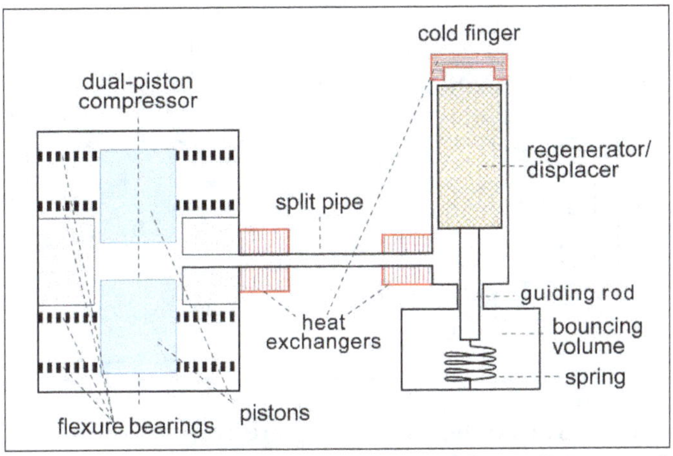

Schematic diagram of a split-pair Stirling refrigerator. The cooling power is supplied to the heat exchanger of the cold finger. Usually the heat flows are so small that there is no need for physical heat exchangers around the split pipe.

Another type of Stirling cooler is the split-pair type, consisting of a compressor, a split pipe, and a cold finger. Usually there are two pistons moving in opposite directions driven by AC magnetic fields (as in loudspeakers). The pistons can be suspended by so-called flexure bearings. They provide stiffness in the radial direction and flexibility in the axial direction. The pistons and the compressor casing don't touch so no lubricants are needed and there is no wear. The regenerator in the cold finger is suspended by a spring. The cooler operates at a frequency near the resonance frequency of the mass-spring system of the cold finger.

GM-refrigerators

Gifford-McMahon (GM) coolers have found widespread application in many low-temperature systems e.g. in MRI and cryopumps. Fig. is a schematic diagram. Helium at pressures in the 10 to 30 bar range is the working fluid. The cold head contains a compression and expansion space, a regenerator, and a displacer. Usually the regenerator and the displacer are combined in one body. The pressure variations in the cold head are obtained by connecting it periodically to the high- and low-pressure sides of a compressor by a rotating valve. Its position is synchronized with the motion of the displacer. During the opening and closing of the valves irreversible processes take place, so GM-coolers have intrinsic losses. This is a clear disadvantage of this type of cooler. The advantage is that the cycle frequencies of the compressor and the displacer are uncoupled so that the compressor can run at power-line frequency (50 or 60 Hz) while the cycle of the cold head is 1 Hz. In this way the swept volume of the compressor can be 50(60) times smaller than of the cooler. Basically (cheap) compressors of domestic refrigerators can be used, but one must prevent overheating of the compressor as it is not designed for helium. One must also prevent oil vapor from entering the regenerator by high-quality purification traps.

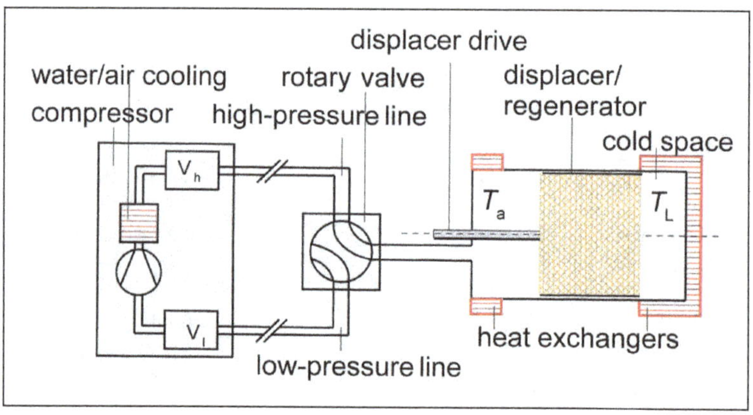

Schematic diagram of a GM-cooler. V_l and V_h are buffer volumes of the compressor. The compression heat is removed by the cooling water of the compressor via a heat exchanger. The rotary valves alternatingly connect the cooler to the high- and the low-pressure sides of the compressor and runs synchronous with the displacer.

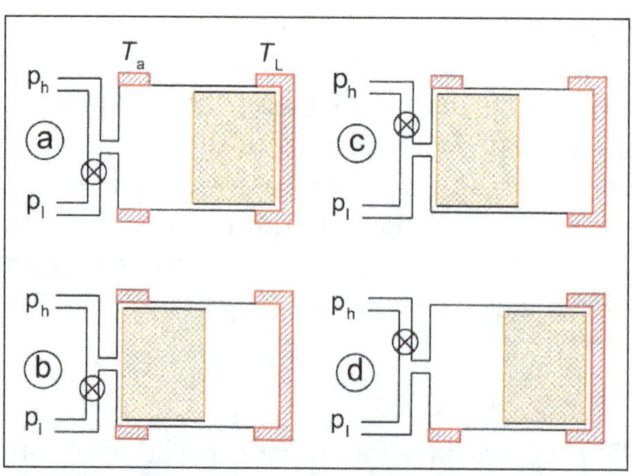

The four stages in the cooling cycle of the GM cooler.

The cycle can be divided in four steps, with the figure above as follows:

The cycle starts with the low-pressure (lp) valve closed, the high-pressure (hp) valve open, and the displacer all the way to the right (so in the cold region). All the gas is at room temperature.

- From a to b. The displacer moves to the left while the cold head is connected to the hp side of the compressor. The gas passes the regenerator entering the regenerator at ambient temperature T_a and leaving it with temperature T_L. Heat is released by the gas to the regenerator material.
- From b to c. The hp valve is closed and the lp valve opened with fixed position of the displacer. Part of the gas flows through the regenerator to the lp side of the compressor. The gas expands. The expansion is isothermal so heat is taken up from the application. This is where the useful cooling power is produced.
- From c to d. The displacer moves to the right with the cold head connected to the lp side of the compressor forcing the cold gas to pass the regenerator, while taking up heat from the regenerator.
- From d to a. The lp valve is closed and the hp valve opened with fixed position of the displacer. The gas, now in the hot end of the cold head, is compressed and heat is released to the surroundings. In the end of this step we are back in position a.

Pulse-tube Refrigerators

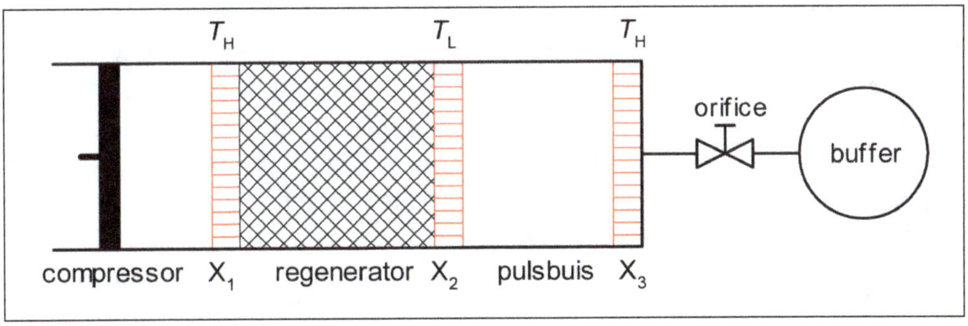

Schematic diagram of a Stirling-type single-orifice PTR.

For completeness a so-called Stirling-type single-orifice PTR is represented schematically in the figure above. From left to right it consists of: a piston which moves back and forth; a heat exchanger X_1 (after cooler) where heat is released at room temperature (T_a) to cooling water or to the surroundings; a regenerator; a heat exchanger X_L at low temperature (T_L) where heat is absorbed from the application; a tube, often called the pulse tube; a heat exchanger X_3 to room temperature (T_a); a flow resistance (orifice); a buffer volume, in which the pressure p_B is practically constant.

Joule-Thomson Cooler

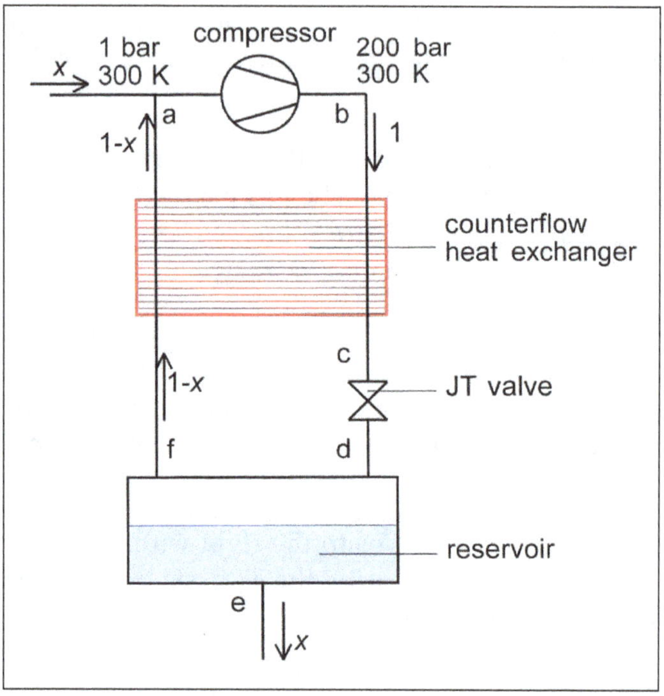

Schematic diagram of a JT liquefier. A fraction *x* of the compressed gas is removed as liquid. At room temperature it is supplied as gas at 1 bar, so that the system is in the steady state.

The Joule-Thomson (JT) cooler was invented by Carl von Linde and William Hampson so it is also called the Linde-Hampson cooler. Basically it is a very simple type of cooler which is widely applied as cryocooler or as the (final stage) of liquefactors. It can easily be miniaturized, but it is also used on a very large scale in the liquefaction of natural gas. A schematic diagram of a JT liquefier, is given in figurw. It consists of a compressor, a counterflow heat exchanger, a JT valve, and a reservoir. In Fig. the pressures and temperatures refer to the case of a nitrogen liquefier. At the inlet of the compressor the gas is at room temperature (300 K) and a pressure of 1 bar (point a). The compression heat is removed by cooling water. After compression the gas temperature is ambient temperature (300 K) and the pressure is 200 bar (point b). Next it enters the warm (high-pressure) side of the counterflow heat exchanger where it is precooled. It leaves the exchanger at point c. After the JT expansion, point d, it has a temperature of 77.36 K and a pressure of 1 bar. The liquid fraction is *x*. The liquid leaves the system at the bottom of the reservoir (point e) and the gas (fraction 1-*x*) flows into the cold (low-pressure) side of the counterflow heat exchanger (point f). It leaves the heat exchanger at room temperature (point a). In order to keep the system in the steady state gas is supplied to compensate for the liquid fraction *x* that has been removed.

When used as a cryocooler it is preferable to use gas mixtures instead of pure nitrogen. In this way the efficiency is improved and the high pressure is much lower than 200 bar.

CRYOGENIC MILLING

Cryogenic milling, or cryo-milling, is the process of using liquid nitrogen to lower the temperate of the material and/or the milling process. cryogenic hammer mills can typically reduce the particle size from small pellets down to the 100s of microns. Cryo-milling can be used for:

- Reducing the material temperature below its glass transition temperature (T_g), which is the point at which a polymer goes from a brittle, glassy state (low temperature) to an elastic, rubbery state (higher temperature). Materials below their T_g are generally friable to some degree.
- Keeping the mill and material from overheating. Hard to mill materials generate a lot of heat during milling, and materials must stay below their melt temps or softening points to be friable.
- Temperature control to prevent hazardous decomposition in the high energy environment of the mill.
- Oxygen (or other gases) exclusion to prevent degradation of the milled material when new surfaces are formed.

The milling of polymers is usually difficult without the use of some liquid nitrogen since many polymers melt in the high energy environment of the mill, especially after a mill has warmed up to its running temperature. The biggest reason, however, is for keeping polymers below their T_g. Many times, being below the T_g is sufficient to render polymers friable. In some cases, the polymers may need further pre-processing in the form of irradiation or solvent crazing.

Water-sensitive materials can be cryo-milled because when liquid nitrogen expands from its liquid state to its gaseous state, its volume increases 694 times. This is enough to quickly exclude any other gas, including water vapor, from being in the atmosphere of the mill, as long as the liquid nitrogen is still flowing. Special care must be taken with any milled material that may come out cold enough to condense water on the surface; sealing up the milled material immediately after completion of milling is usually sufficient.

CRYOGENIC PROCESSOR

A Cryogenic processor is a unit designed to reach ultra-low temperatures (usually around −300 °F / −150 °C) at a slow rate in order to prevent thermal shock to the components being treated. The first commercial unit was developed by Ed Busch in the late 1960s. The development of programmable microprocessor controls allowed the machines to follow temperature profiles that greatly increased the effectiveness of the process. Some manufacturers make cryoprocessors with home computers to define the temperature profile.

Before programmable controls were added to control cryogenic processors, the "treatment" process of an object was previously done manually by immersing the object in liquid nitrogen. This normally caused thermal shock to occur within an object, resulting in cracks to the structure. Modern cryogenic processors measure changes in temperature and adjust the input of liquid nitrogen accordingly to ensure that only small fractional changes in temperature occur over a long period of time. Their temperature measurements and adjustments are condensed into "profiles" that are used to repeat the process in a certain way when treating for similarly grouped objects.

The general processing cycle for modern cryogenic processors occurs within a three-day time window, with 24 hours to reach the optimal bottom temperature for a product, 24 hours to hold at the bottom temperature, and 24 hours to return to room temperature. Depending on the product, some items will be heated in an oven to even higher temperatures. Some processors are capable of providing both the negative and positive extreme temperatures, separate units (a cryogenic processor and a dedicated oven) can sometimes produce better results depending upon the application.

The optimal bottom temperatures for objects, as well as the hold times involved, are determined utilizing a number of different research methods and backed by experience and analysis to determine what works the best for a given product. As new metals are used in different combinations for newer products on the market, processing profiles change to accommodate the changes in structure. Also, profiles will sometimes undergo change from the results of a case study brought to attention by a large manufacturer or consumer of cryogenic services. Generally when a manufacturer sells a cryogenic processor they include the profiles for only that year of manufacture, or, more typically, profiles from when the processor model was first engineered, which sometimes will date back several years. Many businesses will include outdated profiles simply because they do not have adequate funding to perform the necessary ongoing research.

For people looking to find thermal profiles for cryogenics, a number of companies maintain thermal profiles of various products that are updated for accuracy at least a few times a year with their ongoing research, including data from independent trials and studies. However, obtaining these profiles is sometimes difficult if they are not used for educational purposes (mainly institutional research), as they typically only provide the updated profiles to their longtime "service center" partners around the world.

Overall, cryogenic processors are radically changing the way that cryogenics used to be done. Many years ago, cryogenics was simply theoretical, with spotty results when there were improvements. Now, cryogenic processors are ensuring the accurate and consistent results for all products that are treated these days. As the technology sector improves, cryogenic processors will only get better as they benefit from new computer systems. Ongoing research in the future will also improve their temperature treatment profiles.

COOLING BATH

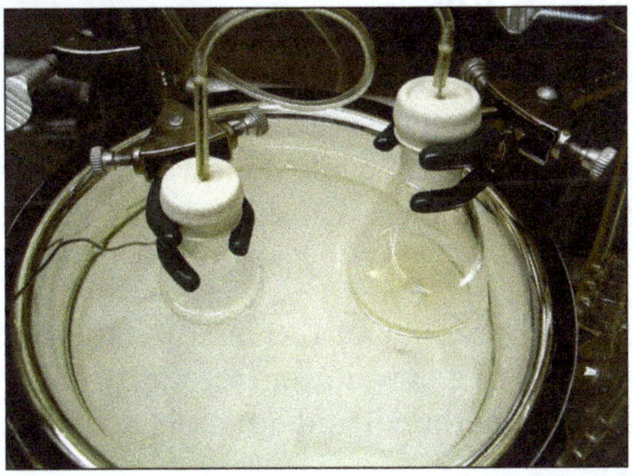

A typical experimental setup for an aldol reaction. Both flasks are submerged in a dry ice/acetone cooling bath (−78 °C) the temperature of which is being monitored by a thermocouple (the wire on the left).

A cooling bath, in laboratory chemistry practice, is a liquid mixture which is used to maintain low temperatures, typically between 13 °C and −196 °C. These low temperatures are used to collect liquids after distillation, to remove solvents using a rotary evaporator, or to perform a chemical reaction below room temperature.

Cooling baths are generally one of two types: (a) a cold fluid (particularly liquid nitrogen, water, or even air) — but most commonly the term refers to (b) a mixture of 3 components: (1) a cooling agent (such as dry ice or water ice); (2) a liquid 'carrier' (such as liquid water, ethylene glycol, acetone, etc.), which transfers heat between the bath and the vessel; ; and (3) an additive to depress the melting-point of the solid/liquid system.

A familiar example of this is the use of an ice/rock-salt mixture to freeze ice cream. Adding salt lowers the freezing temperature of water, lowering the minimum temperature attainable with only ice.

Mixed solvent cooling baths (% by volume)			
% Glycol in EtOH	Temp (°C)	% H_2O in MeOH	Temp (°C)
0%	-78	0%	−97.6
10%	-76	14%	-128
20%	-72	20%	N/A
30%	-66	30%	−72

40%	-60	40%	−64
50%	-52	50%	−47
60%	-41	60%	−36
70%	-32	70%	−20
80%	-28	80%	−12.5
90%	-21	90%	−5.5
100%	-17	100%	0

Mixed Solvent Cooling Baths

Mixing solvents creates cooling baths with variable freezing points. Temperatures between approximately −78 °C and −17 °C can be maintained by placing coolant into a mixture of ethylene glycol and ethanol, while mixtures of methanol and water span the −128 °C to 0 °C temperature range., Dry ice sublimes at −78 °C, while liquid nitrogen is used for colder baths.

As water or ethylene glycol freeze out of the mixture the concentration of ethanol/methanol increases. This leads to a new, lower freezing point. With dry ice these baths will never freeze solid, as pure methanol and ethanol both freeze below −78 °C (-98 °C and -114 °C, respectively).

Relative to traditional cooling baths, solvent mixtures are adaptable for a wide temperature range. In addition, the solvents necessary are cheaper and less toxic than those used in traditional baths.

Traditional Cooling Baths

Traditional cooling bath mixtures		
Cooling agent	Organic solvent or salt	Temp (°C)
Dry ice	p-xylene	+13
Dry ice	Dioxane	+12
Dry ice	Cyclohexane	+6
Dry ice	Benzene	+5
Dry ice	Formamide	+2
Ice	Salts	0 to −40
Liquid N_2	Cycloheptane	−12
Dry ice	Benzyl alcohol	−15
Dry ice	Tetrachloroethylene	−22
Dry ice	Carbon tetrachloride	−23
Dry ice	1,3-Dichlorobenzene	−25
Dry ice	o-Xylene	−29
Dry ice	m-Toluidine	−32

Dry ice	Acetonitrile	−41
Dry ice	Pyridine	−42
Dry ice	m-Xylene	−47
Dry ice	n-Octane	−56
Dry ice	Isopropyl ether	−60
Dry ice	Acetone	−78
Liquid N_2	Ethyl acetate	−84
Liquid N_2	n-Butanol	−89
Liquid N_2	Hexane	−94
Liquid N_2	Acetone	−94
Liquid N_2	Toluene	−95
Liquid N_2	Methanol	−98
Liquid N_2	Cyclohexene	−104
Liquid N_2	Ethanol	−116
Liquid N_2	n-Pentane	−131
Liquid N_2	Isopentane	−160
Liquid N_2	(none)	−196

Water and Ice Baths

A bath of ice and water will maintain a temperature 0 °C since the melting point of water is 0 °C. However, adding a salt such as sodium chloride will lower the temperature through the property of Freezing-point depression. Although the exact temperature can be hard to control, the weight ratio of salt to ice influences the temperature:

- −10 °C can be achieved with a 1 to 2.5 ratio by weight of calcium chloride hexahydrate to ice.
- −20 °C can be achieved with a 1 to 3 ratio by weight of sodium chloride to ice.
- −40 °C can be achieved with a 1 to 0.8 ratio by weight of calcium chloride hexahydrate to ice.

Dry Ice Baths at −78 °C

Since dry ice will sublime at −78 °C, a mixture such as acetone/dry ice will maintain −78 °C. Also, the solution will not freeze because acetone requires a temperature of about −93 °C to begin freezing. Therefore, other liquids with a lower freezing point (pentane: −95 °C, isopropyl alcohol: −89 °C) can also be used to maintain the bath at −78 °C.

Dry Ice Baths Above −77 °C

In order to maintain temperatures above −77 °C, a solvent with a freezing point above −77 °C must be used. When dry ice is added to acetonitrile then the bath will begin cooling. Once the temperature reaches −41 °C, the acetonitrile will freeze. Therefore, dry ice must be added slowly to avoid freezing the entire mixture. In these cases, a bath temperature of −55 °C can be achieved by choosing a solvent with a similar freezing point (n-octane freezes at −56 °C).

Liquid Nitrogen Baths Above −196 °C

Liquid nitrogen baths follow the same idea as dry ice baths. A temperature of −115 °C can be maintained by slowly adding liquid nitrogen to the organic solvent (ethanol) until it begins to freeze (ethanol freezes at −116 °C).

Water/Ice Alternatives

In water and ice-based baths, tap water is commonly used due to ease of access and the higher costs of using ultrapure water. However, tap water and ice derived from tap water can be a contaminant to biological and chemical samples. This has created a host of insulated devices aimed at creating a similar cooling or freezing effect as ice baths without the use of water or ice.

Safety Recommendations

The American Chemical Society notes that the ideal organic solvents to use in a cooling baths have the following characteristics: 1. Nontoxic vapors 2. Low viscosity 3. Non-flammability 4. Low volatility 5. Suitable freezing point. In some cases, a simple substitution can give nearly identical results while lowering risks. For example, using dry ice in 2-propanol rather than acetone yields a nearly identical temperature but avoids the volatility of acetone.

HELIUM CRYOGENICS

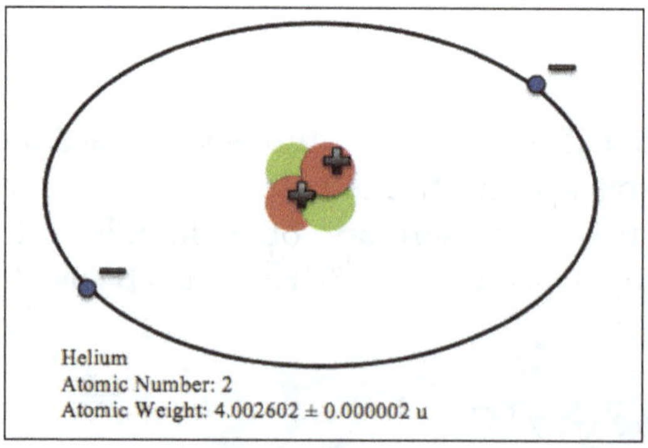

In the field of cryogenics, helium is utilized for a variety of reasons. The combination of helium's extremely low molecular weight and weak interatomic reactions yield interesting properties when helium is cooled below its critical temperature of 5.2 K to form a liquid. Even at absolute zero (0K), helium does not condense to form a solid. In this state, the zero point vibrational energies of helium are comparable to very weak interatomic binding interactions, thus preventing lattice formation and giving helium its fluid characteristics. Within this liquid state, helium has two phases referred to as helium I and helium II. Helium I displays thermodynamic and hydrodynamic properties of classical fluids, along with quantum characteristics. However, below its lambda point of 2.17 K, helium transitions to He II and becomes a quantum superfluid with zero viscosity.

Under extreme conditions such as when cooled beyond T_λ, helium has the ability to form a new state of matter, known as a Bose–Einstein condensate (BEC), in which the atoms virtually lose all their energy. Without energy to transfer between molecules, the atoms begin to aggregate creating an area of equivalent density and energy. From observations, liquid helium only exhibits super-fluidity because it contains isolated islands of BECs, which have well-defined magnitude and phase, as well as well-defined phonon–roton (P-R) modes. A phonon refers to a quantum of energy associated with a compressional wave such as the vibration of a crystal lattice while a roton refers to an elementary excitation in superfluid helium. In the BEC's, the P-R modes have the same energy, which explains the zero point vibrational energies of helium in preventing lattice formation.

When helium is below T_λ, the surface of the liquid becomes smoother, indicating the transition from liquid to superfluid. Experiments involving neutron bombardment correlate with the existence of BEC's, thereby confirming the source of liquid helium's unique properties such as super-fluidity and heat transfer.

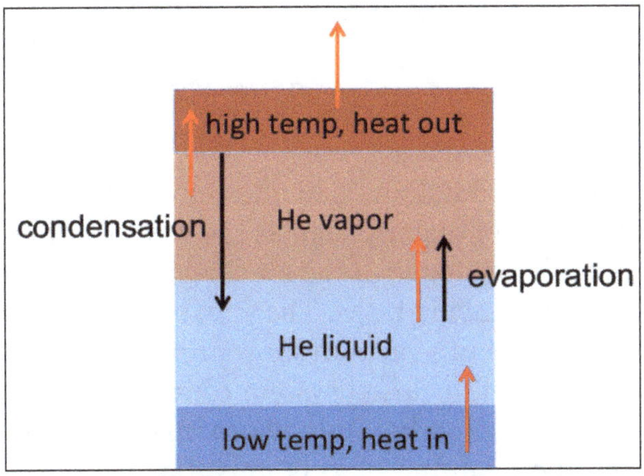

A schematic of a helium cooling system; heat flow is represented by red arrows and helium flow is by black arrows.

Though seemingly paradoxical, cryogenic helium systems can move heat from an area of relatively low temperature to an area of relatively high temperature. Though this phenomenon appears to violate the second law of thermodynamics, experiments have shown this to prevail in systems where the area of low temperature is constantly heated, and the area of high temperature is constantly cooled. It is believed this phenomenon is related to the heat associated with the phase change between liquid and gaseous helium.

Applications

Superconductors

Liquid helium is used as a coolant for various superconducting applications. Notable are particle accelerators where magnets are used for steering charged particles. If large magnetic fields are required then superconducting magnets are used. In order for superconductors to be efficient, they must be kept below their respective critical temperature. This requires very efficient heat transfer. Because of the reasons discussed previously, superfluid helium can be used to effectively transfer heat away from superconductors.

Quantum Computing

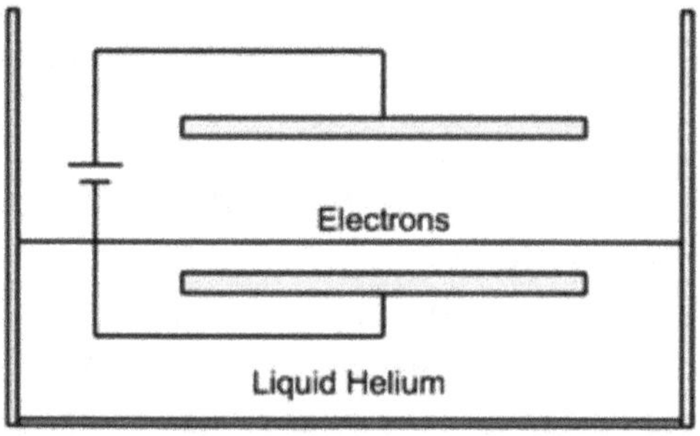

A capacitor inserted with the electron system on the surface of liquid helium.

One proposed use for superfluid helium is in quantum computing. Quantum computers utilize the quantum states of matter, such as the electron spin, as individual quantum bits (qubits), a quantum analogue of the bit used in traditional computers to store information and perform processing tasks. The spin states of the electrons present on the surface of superfluid helium in a vacuum show promise as excellent qubits. In order to be considered a usable qubit, a closed system of individual quantum objects must be created that interact with each other, but whose interaction with the outside world is minimal. In addition, the quantum objects must be able to be manipulated by the computer, and the quantum system's properties must be readable by the computer to signal the termination of a computational function. It is believed that in vacuum, superfluid helium satisfies many of these criteria since a closed system of its electrons can be read and easily manipulated by the computer in a similar fashion as electrostatically manipulated electrons in semiconductor heterostructures. Another beneficial aspect of the liquid helium quantum system is that application of an electrical potential to liquid helium in a vacuum can move qubits with little decoherence. In other words, voltage can manipulate qubits with little effect on the ordering of the phase angles in the wave functions between the components of the liquid helium quantum system.

X-ray Crystallography

The advent of high-flux X-rays provides a useful tool for developing high-resolution structures of proteins. However, higher energy crystallography incurs radiation damage to the proteins studied. Cryogenic helium systems can be used with greater efficacy than nitrogen cryogenic systems to prevent radical damage to protein crystals.

CRYOGENIC SEAL

Cryogenic seals provide a mechanical containment mechanism for materials held at cryogenic temperatures, such as cryogenic fluids. Various techniques, including soldering and welding are available for creating seals; however, specialized materials and processes are necessary to hermetically entrap cryogenic constituents under vacuum-tight conditions. Most commonly used are liquid helium and liquid nitrogen, which boil at very low temperatures, below −153 °C (120 K), as well as hydrocarbons with low freezing points and refrigerating mixtures. Pure indium wire or solder preform washers are accepted as the most reliable low temperature sealing materials. When correctly formed, indium will afford leak rates of less than 4.0x10 -9 mbar- liter/sec. Alternative cryogenic seal materials include silicone grease conical seals, and Pb/Sn (lead-tin) wire seals.

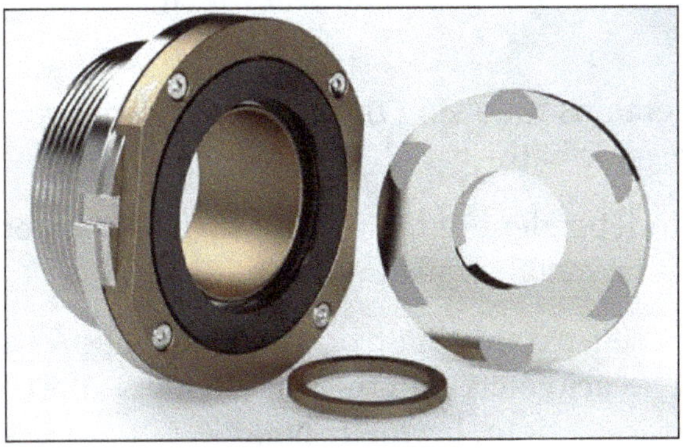

Applications

Applications which utilize cryogenic seals include:

- Magnetic resonance imaging (MRI),
- Chromatography,

- Dilution Refrigeration Units,
- Cooled Detectors,
- Optical Windows,
- Infrared Detectors,
- Centrifugal Cryogenic Pumps,
- Unmanned Aerial Vehicle Systems,
- Missile Warning Receivers,
- Satellite Tracking Systems,
- Infrared Telescopes.

Indium Seals

Advantages

Advantages of indium cryogenic seals:

- Established/proven design techniques for indium seal assembly,
- Option for disassembly and re-assembly,
- Indium can be reformed into useful seals after use,
- Soft and pliable at room temperature, due to the low melting temperature of indium, so it fills imperfections. This creates an impervious bond between the mating surfaces, crafting a hermetic seal which remains malleable at cryogenic temperatures,
- Seal integrity remains following thermal shock from room temperature to immersion in cryogenic bath,
- Seal quality is independent of the mating surface composition, for instance ceramic, germanium, metal, or glass,
- Indium forms a self-passivating oxide layer, 80-100Å thick. This layer is easy to remove with an acid etch, and the underlying, exposed indium metal can be compressed to form a tight, hermetic bond.

Disadvantages

- Bulky mechanical structure required to compress indium between the flanges,
- Pulsating loads cause creep of indium seals, which loosens the bolt tension, thereby reducing the quality of the seal.

Process Information for Indium Seals

- Mating surfaces should be kept as clean as possible, and may be cleaned using acetone.
- Clean, oxide-free indium will cold weld to itself. The mating ends of a wire seal will weld together under compression.
- A more reliable alternative to a seal made from indium wire is a seal that uses an indium washer. Washers minimize the risk of seal degradation and cryogenic leaks by eliminating the interface between connected butt ends of wire. Washers are manufactured as a continuous ring with no breaks.
- As many fasteners as possible should be used to clamp the indium seal.
- Indium material used must be ultra-pure (99.9 minimum purity) to prevent hardening of the material at sub-zero temperatures, as well as to restrict impurities of elements with low vapor-pressure.
- Material used for indium cryogenic seals should be manufactured from vacuum-cast material to prevent outgassing after fixturing in the assembly.

Reliability Testing

- Helium leak tests,
- Cryogenic temperature shock testing.

Types

- Compact indium seal,
- Compressible hermetic seal,
- Compression seal,
- Cryogenic vacuum seal,
- Demountable cryogenic seal,
- Indium cryogenic vacuum seal,
- Indium o-ring flange seals,
- Indium seal,
- Indium wire o-ring,
- Indium wire seal,

- Low profile indium seal,
- Low temperature seal,
- Reusable cryogenic vacuum seal,
- Reusable indium wire seal,
- Reusable, low-profile, cryogenic wire seal,
- Soft metal seal,
- Vacuum compatible seal,
- Vacuum compatible seal at cryogenic temperature,
- Corner joint seal,
- Face joint seal.

CRYOTANK

Cryotanks at the Institute of Plant Industry, Saint Petersburg, Russia.

Cryotank or cryogenic tank is a tank that is used to store frozen biological material.

The term "cryotank" refers to storage of super-cold fuels, such as liquid oxygen and liquid hydrogen. Cryotanks and cryogenics can be seen in many sci-fi movies, but they are still currently undeveloped. All that needs to be done is for a human to be loaded into the tank and then they can be frozen until a time comes when any diseases they have can be cured and they can live an even longer life. This could also be used in space travel and just preserving human life in general. The problem with this is when the human body is frozen, ice crystals form in the cells. The ice crystals then continue to expand rupturing the cell wall and destroying the integrity of the cell, or killing it.

This means in order for humans to undergo the cryogenic process a way to significantly raise the levels of glucose produced in the human body is needed.

Material uses

Cryogenic tanks are used to store natural gases such as oxygen, argon, nitrogen, helium, hydrogen and other materials. Tanks can store the materials at the correct temperature and pressure for transportation.

Application of Carbon Fiber Composites in Cryotank

Carbon fiber reinforced resin matrix composite materials (CFRC) are being used in the aerospace industry as a means of reducing vehicle weight. CFRC have advantages in high strength-to-weight and high stiffness-to-weight ratios. For future heavy lift launch vehicles and space exploration structures, advanced lightweight composites will be fully utilized in order to minimize vehicle weight, and CFRC in space applications requires rigorous development to demonstrate robustness, durability, and high factors of safety.

The future heavy lift launch vehicles require extremely high propellant mass fractions to achieve the designed performance. This drives the designers to incorporate lightweight materials in as many structures as possible. Propellant fuel tanks account for a large proportion in the launch vehicles, both of structural mass and geometric space. Approximately 60% of the dry mass of a launch vehicle is the fuel and oxidizer tanks. The implementation of composite cryogenic propellant fuel tanks (cryotank) for future heavy lift launch vehicles could greatly reduce the vehicle's weight by replacing the identically sized cryotanks constructed of metallic materials. United States' Committee on Materials Needs and R&D Strategy for Future Military Aerospace Propulsion Systems reported that composites offer the potential for the greatest mass reduction of all of the materials for tank.

For the case of Delta IV heavy lift launch vehicle, as shown in Figure , compared to Li-Al fuel tank, the weight saving of upperstage composite cryotanks were 43 and 26%, respectively.

Delta IV heavy.

In addition, composite design could reduce fabrication cost. Delta II faring, Delta III faring, and interstage production data have shown that composite launch vehicle structures are less expensive than metal ones.

Graphite-epoxy composite cryogenic tank development began at Boeing (then McDonnell Douglas) in 1987 and continues today, primarily for reusable launch vehicles (RLV) and heavy lift vehicles.

LH_2 Cryotank

The cryogenic tanks are the dominating components of the vehicle structure. To achieve weight reduction of the next-generation launch vehicles, carbon fiber reinforced polymeric-based composites are being explored for cryogenic liquid fuel tank. A composite cryotank structure can save 30% by weight than lithium aluminum alloy.

Cryogenic composite tank development began in 1987. First, most of the effort has been devoted to liquid hydrogen tanks mainly because liquid hydrogen tanks are larger than liquid oxygen tanks, lightweight materials would provide a proportionately greater weight reduction. The other reason was a common assumption that organic composite materials could not be used in LOX tanks.

NASA has been exploring advanced composite materials and processes to reduce the overall cost and weight of liquid hydrogen (LH_2) cryotanks while maintaining the reliability of existing designs.

Composite liquid hydrogen tank development in NASA went through the National Aerospace Plane (NASP) program, the Single-Stage-to-Orbit (SSTO) vehicle X-33 program, and the recent second generation RLV applications under the Space Launch Initiative (SLI) and the Composite Cryotank Technology Demonstration Project (CCTD) (2011–2014). The fundamental issues for composite liquid hydrogen tanks are the hydrogen permeability and cryogenic mechanical properties of the composite material.

DC-XA Composite LH_2 Cryotank

Liquid hydrogen is an essential but highly volatile fuel used in launch vehicles. It must be stored and used at −253 °C, a temperature that causes most materials to become quite brittle. Liquid hydrogen also has an extremely fine molecular structure, which allows it to seep through the tiniest gaps. Therefore, liquid hydrogen tanks present an extreme challenge in engineering materials because of hydrogen permeability and cryogenic properties.

Work in composite cryogenic tank development at McDonnell Douglas (MDA) began in 1987. MDA solved the problem of hydrogen permeation and cryogenic properties of composites successfully, and the composite LH_2 tank for DC-XA was designed, developed, and fabricated. The tank was 8 feet (2.43 m) in diameter, 16 feet long (4.88 m), constructed with IM7/8552 toughened epoxy material from Hercules. Automated fiber placement (AFP) was employed to manufacture the cryotank. The tank was incorporated with lightweight internal insulation modeled after the Saturn S-IVB design. The composite tank was 33% lighter than the DC-X tank.

DC-XA provided the first flight tests in real-world operating environments for composite liquid hydrogen tank in 1996, as shown in the figure below. The tank passed the ground and flight tests successfully. It experienced approximately 55 pressure cycles throughout the ground and flight test program, demonstrated the acceptability of composite liquid hydrogen tank for future launch vehicle.

DC-XA composite LH_2 cryotank.

X-33 Composite LH_2 Cryotank

Lockheed Martin Space Systems Company (LM) is one of the world leaders in large cryogenic tank technology. The composite LH_2 cryogenic tank for X-33 vehicle was designed, developed, and built by LM and test by NASA. The tank was a conformal, load-bearing, composite sandwich structure, consisting of an outer facesheet, honeycomb core, and inner facesheet as shown in the figure below. The inner and outer facesheets were constructed with IM7/977-2 and the core was constructed with KorexT-M+3-pcf honeycomb. The figure shows the overall dimensions of the tank. The tank is 28.5 feet in length, 20.0 feet in width, and 14.0 feet in height.

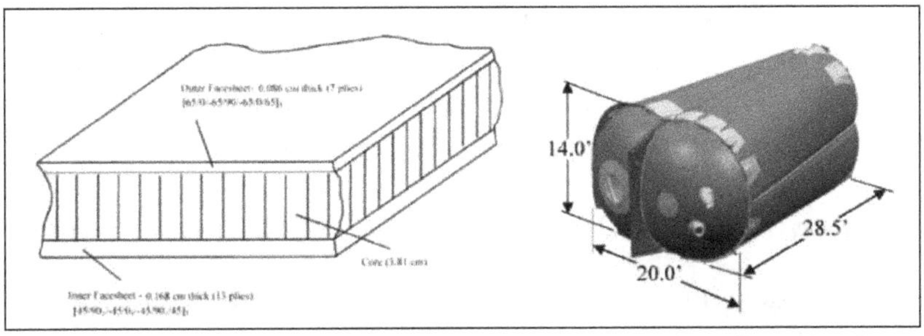

Lockheed X-33 composite LH_2 cryotank (a) configuration of honeycomb sandwich structure and (b) tank dimensions.

The sandwich LH_2 tank of the X-33 Demonstrator was ground-tested at Marshall Space Flight Center on November 3, 1999. It failed this validation test when the outer skin and core of the sandwich separated from the inner skin, although the inner skin was undamaged, several fractures were observed in the outer skin in the same lobe.

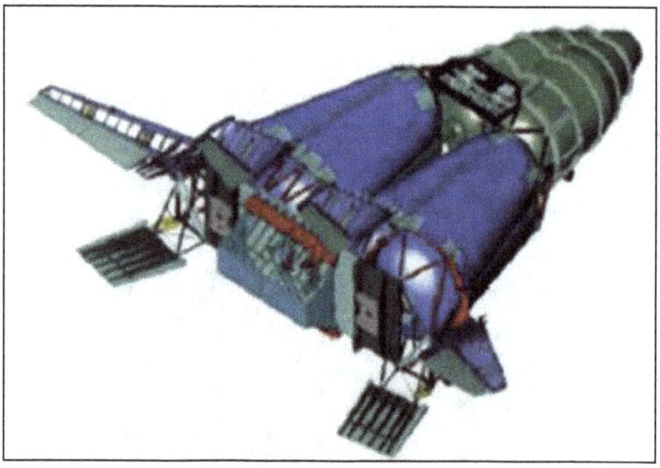

X-33 vehicle.

After extensive study, NASA determined that the delaminations were likely caused by a combination of factors. The most probable cause of the failure was determined to be a combination of the following phenomena:

- Microcracking of the inner facesheet with gaseous hydrogen (GH_2) infiltration,
- Cryopumping of the exterior nitrogen (N_2) purge gas,
- Reduced bondline strength and toughness,
- Manufacturing flaws and defects,
- Infiltration of GH_2 into the core, which produced higher than expected core pressures.

This setback did not discourage NASA to further develop composite cryotanks. The X-33 project was the first of its kind in large-scale composite cryotanks. The tank design was very novel with a number of cutting edge technologies. The delamination failure during the testing provided a valuable case study for NASA and Lockhed Martin to continue the composite material application in cryotank.

Space Launch Initiative (SLI) Composite Cryotank Program

Northrop Grumman designed, developed, and fabricated a cylindrical composite LH_2 cryotank for Space Launch Initiative Composite Cryotank Program, as shown in the figure below. The tank was constructed as a sandwich structure with carbon fiber composite skins sandwiched between nonmetallic honeycomb core. By using thinner carbon fiber laminars and increased number of cross laminates, the microcracks in the com-posite skins were significantly reduced by a factor of 16.

North Grumman's LH_2 cryotank.

Other major innovations in this program included the multifunctional sandwich core and the out-of-autoclave process. Taking the advantage of the sandwich structure, hydrogen permeation was stopped by a barrier of aluminum foil between the inner skin and the honeycomb core. Any hydrogen leakage was vacuumed out through the perforated honeycomb core. In addition, the core acted as a thermal insulation layer. The cryotank manufacturing avoided the expensive autoclave by the novel ultrasonic tape lamination approach. The ultrasonic energy provided excellent compaction while depositing the carbon fiber tapes. The final composite structure was cured in an oven at ambient pressure.

The testing involved the simulated load cases close to a rocket launch. The composite cryotank, in 1.8 m diameter and 4.5 m long, was filled with liquid hydrogen. The tank was also subject to an internal pressure of 827 KPa and an axial load along the vertical axis of a launch vehicle. The composite tank was ¼ scaled size of a typical reusable launch vehicle cryotank. Further tests were performed to fill, apply internal and external loads, and drain the tank repeatedly 40 times, in order to study its structural integrity under cryogenic temperatures and its reusability.

The Northrop Grumman/NASA team proved that a cryogenic fuel tank made from composite materials has the structural integrity to withstand the mechanical and thermal stresses associated with repeated fueling and simulated launch cycles.

CCTD Composite Cryotank

The Composite Cryotank Technology Demonstration Project (CCTD) was part of the Space Technology Program and the Game Changing Development (GCD) Program for

NASA. The program had a clear focus to deliver the proven technologies that would be deployed in future space flight demonstration. An extensive full-scale ground-based testing and lab experimentations were implemented. The development and demonstrations outlined in the CCDT Program were based on the relevant aerospace industrial experience over the last 20 years. The project successfully designed, manufactured, and tested a full-composite LH_2 cryotank. The composite cryotank reportedly achieved 30% saving in weight and 25% in cost, compared with a baseline aluminum alloy cryotank.

NASA worked with four competing industry partners—ATK, Boeing, Lockheed Martin, and Northrop Grumman at the design phase of the project with the aim to reduce weight by 30%, reduce cost by 25%, and withstand a strain of 0.005. Four competing conceptual designs differing in materials, structures, manufacturing processes were assessed through coupon testing and finite element stress analysis.

A sample of the design requirements is shown in the figure below: an inner tank that could contain pressurized LH_2 at cryogenic temperatures, fitted with an outer skirt that could take the axial compression loads during launch-vehicle takeoff. All of the phase I concepts were required to use Government furnished information (GFI) IM7/977-2 lamina material property.

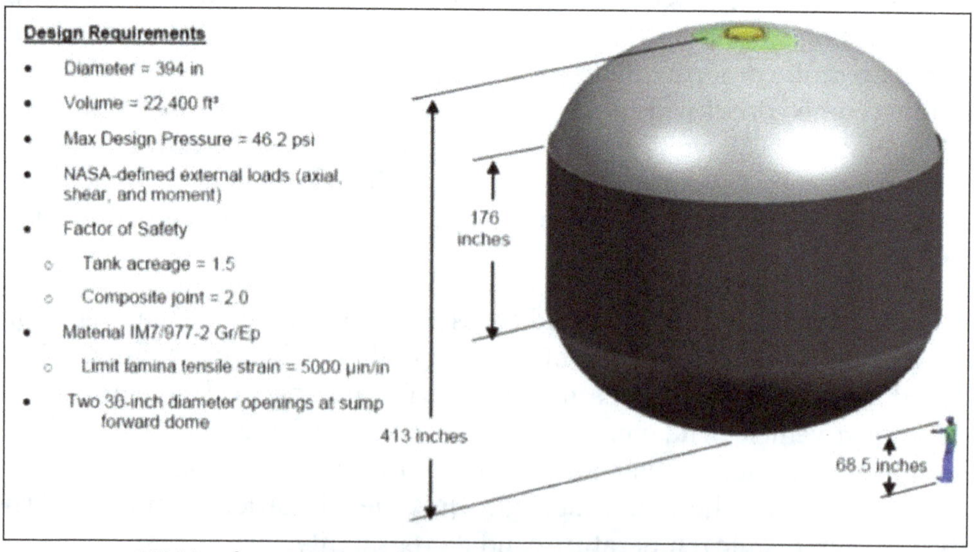

NASA reference Cryotank geometry and high level requirements.

The NASA team developed a metallic aluminum alloy cryotank concept for comparison to three industry IM7/977-2 composite concepts with the same overall dimensions: Boeing fluted core, Lockheed-Martin externally stiffened, and Northrop Grumman sandwich, as shown in Figure . The weight comparison for the four concepts is shown in Table . All three composite concepts exceeded the 30% weight reductions also had designs where the laminates were at or under the 5000 με strain limit desired by the CCTD Project when compared to the metallic cryotank. All of the three tanks were fabricated by automated fiber placement process. A cost analysis effort showed that 20–25% cost saving can be achieved by utilizing AFP process.

Company structural concept	Weight/lb	Weight saving/%
NASA TRL9 metallic baseline	10,925	–
Boeing fluted core	6696	38.7
Lockheed External Box Stiffened	6572	39.8
NGC Honeycomb Sandwich	6252	42.8

Table: Comparison of the three industry composite designs to the metallic design from NASA.

The results further showed that thin plies (65 or 70 g/m^2) are effective in resisting microcracks and thereby minimizing LH_2 permeation.

Boeing's design and analyses showed that when designing to a 5000 με limit strain level, a 39% weight saving over a comparable aluminum-lithium tank designed using mature materials and manufacturing techniques can be realized. A cost analysis effort showed that 20–25% cost saving can be achieved by utilizing an automated fiber placement process.

Phase II of CCTD involved the design, analysis, fabrication, and testing of large scale (2.4-m diameter precursor and 5.5-m diameter demonstrator) composite cryotanks. The two tanks incorporated the design features and strain levels that represent a full-scale (8.4-m diameter) Space Launch System (SLS) propellant tank. Design features included a one-piece wall design that minimized tank weight.

Both tanks were fabricated at Boeing using automated fiber placement on breakdown tooling. The tanks were made of Cytec's CYCOM 5320-1 out-of-autoclave (OOA) prepreg, hybrid laminate was employed, using a combination of thick plies (145 g/m²), which can be placed relatively quickly and enable large (up to 10-m diameter) cryotank fabrication, and thin plies (70 g/m²), which create a microcrack-resistant laminate that helps prevent hydrogen permeation .

2.4 m diameter tank, a robotic arm applies composite laminate.

The 2.4-m diameter all-composite precursor tank was fitted with a laminate skirt that could take the axial compression loads during launch-vehicle takeoff and tested on June 25, 2013 at NASA MSFC, the tank was successfully pressure tested. The test met all requirements: stepwise fill with liquid hydrogen (−223°C) to 90% volume capacity followed by pressurizing the tank to 931 KPa. The 2.4 m tank was then cycled through 20 pressure/vent cycles, measuring hydrogen gas permeation on the tank dome.

8.4 m diameter tank.

NASA Space Launch System (SLS) has an 8.4-m diameter (8.4 m) Ares Vupper stage. NASA chose a 5.5-m diameter test article for the second phase of the program. Boeing was the only partner to produce and test this size tank with the existing infrastructure. A 5.5-m diameter cryotank was of sufficient scale. To prevent the delaminations occurred in X-33 composite cryotanks, subscale 5.5-m diameter CCTD demonstrator tank was fitted with an innovative fluted-core skirt.

NASA completed a demanding series of tests inside the test stand at MSFC. On March 26, 2014, the 5.5 m cryotank was subjected to flight loads in combination with pressure loads. Structural loads were applied to simulate the stress during a space flight. Liquid hydrogen at 253°C was filled in the composite tank. At the same time, a cyclic pressure of 138 to 207 KPa was applied to the tank.

The 5.5 m tank was pressurized to 400 KPa, reaching a maximum acreage strain of 5136 microstrain and demonstrating safety factors above 1.5 in the scarf joints. At the 5136 test strain, the permeation performance does not meet the CCTD goals, but well within the allowable for an upper stage or boost stage composite tank application.

The hydrogen permeation measured in the CCTD tanks are likely due to the porosity, estimated to be approximately 3%, from the low-pressure curing having facilitated void and crack formation, even in the thin plies.

The project confirmed that composite cryotanks can achieve a 33% weight savings compared to aluminum-lithium cryotanks, and it demonstrated permeation performance that meets the allowable for upper stage and boost stage applications.

This is the first effort to successfully build and test a tank of 5.5-m diameter.

There are several accomplishments in CCTD cryotank of phase II:

5320-1 OoA Epoxy Resin Matrix

Boeing tested a new material-Cycom 5320-1 that does not require expensive autoclave curing. Cycom 5320-1 is a toughened epoxy prepreg resin system designed for out-of-autoclave manufacturing of primary structures. With a lower curing temperature, the resin system is suitable for prototyping where low cost tooling or vacuum-bag-only curing is required.

Cycom 5320-1 handles similarly to standard prepreg. The difference is that the vacuum-bag-only cured composites produce the quality equivalent to the autoclave process, with minimum porosity and competitive mechanical properties. It was received as thin sheets of B-stage film.

Thin Ply Prepreg-70 g/m^2, Hybrid Laminate

To reduce hydrogen permeation levels, hybrid laminate combination of thick plies (145 g/m^2), which can be placed relatively quickly and thin plies (70 g/m^2), which create a microcrack-resistant laminate that helps prevent hydrogen permeation was used in CCTD tanks.

For further evaluation, hybrid laminates fiber-placed panels were produced at Boeing and evaluated at MFSC. The stacking sequence of the laminate was showed in Figure. The laminates were made up of 12 plies of 5.4 mil and 5 plies of 2.5 mil material. The report noted that when thin plies are used and standard laminate consolidation is achieved, permeation performance requirements will be met with very large margins.

Thin and thick ply.

Thin-ply composite structures offer many advantages in composite tank manufacture. They are far more resistant to the formation of microcracks. Also, tougher resins have been developed that offer protection against microcracks (may be used in conjunction with the thin plies). The down-side of the thin plies is that they make manufacturing more difficult. Present development efforts are exploring the use of fiber placement equipment to place the thin plies. Hybrid laminates are demonstrating the same performance as the thin plies. Excellent permeability results are achieved by both methods.

Out-of-autoclave Cure Processing

Autoclave process is the major cost in composite manufacturing. The investment of the large size autoclave oven is formidable for any company. Boeing employed OOA resin matrix, and the tank exhibited approximately 3% porosity.

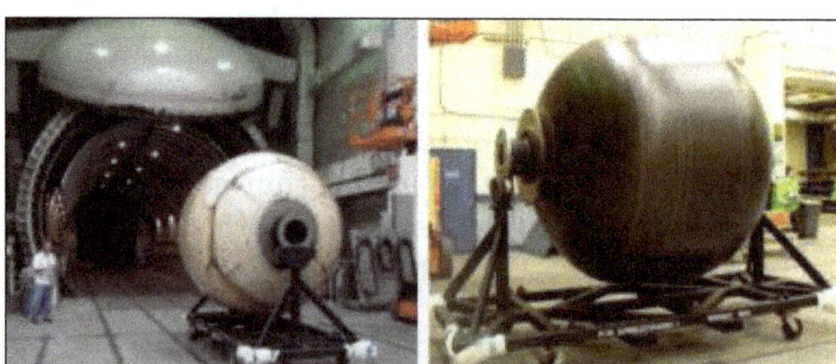

Out-of-autoclave processing.

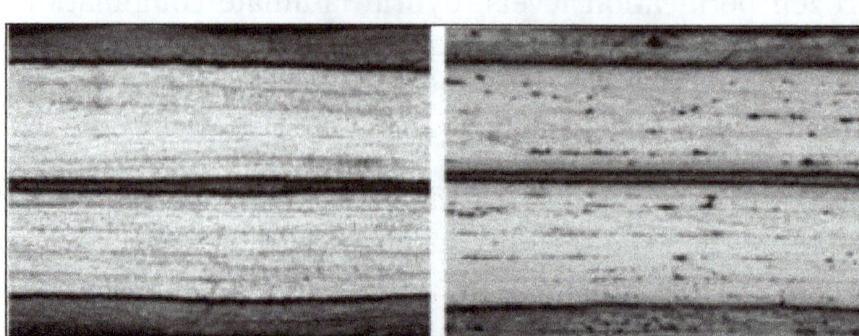

Comparison of autoclave and out-of-autoclave processing.

For further evaluation of OOA processing, two-hybrid laminates fiber-placed panels produced at Boeing and LH_2 tested at MSFC. The laminates were made up of 12 plies of 0.137 mm and 5 plies of 0.064 mm material. The OOA laminates exhibited approximately 4% porosity. Testing with autoclave coupons and the same materials did not show measurable permeability. Evaluation of the OOA laminates revealed that microcrack forms in the thin plies, primarily due to the porosity in the laminate. To deduce porosity, and eliminate permeability for OOA, it is necessary to increase the number of thin plies and to reduce porosity by improving the OOA materials architecture and fiber placement processes.

Robotic AFP

Instead of the gantry-based automated fiber placement (AFP) equipment classically used to build large cylindrical parts, Boeing opted to use robotic fiber placement (AFP). Robotic AFP system enabled improved capabilities–better reach in the dome areas and lower shaft clearance.

Fluted-core Composite Skirt

To prevent the delaminations occurred in X-33 composite cryotanks, Boeing opted an innovative fluted-core skirt, comprising large trapezoidal members-technically, laminate-angled web members with structural-radius fillers between facesheets. The fluted-core skirt was designed to take the compressive load during launch and to vent permeated hydrogen. The skirt structure was co-cured to eliminate potential de-bonding issues at the working temperature of 253°C.

Fluted-core skirt.

Chatt Project

The project Cryogenic Hypersonic Advanced Tank Technologies (CHATT) is a part of the European Commission's Seventh Framework Programme and run on behalf of the Commission by DLR-SART in a multinational collaboration. The main objective was to investigate carbon fiber reinforced composite material for cryogenic fuel tank applications.

Four different composite tanks of CHATT project.

Four different subscale CFRP-tanks have been designed, manufactured, and tested. The CHATT project contributed to significant progress in the design of composite tanks for cryogenic propellant applications in Europe. The European Technology Readiness Level (TRL) of such cryotank is in the range between three and four, while the TRL in the US is considerably more advanced.

LOX Cryotank

The weight reduction provided by composite Lox tank is too great to disregard. The most difficult issue for composite Lox tanks is material compatibility. After solving the problems of the hydrogen permeation and cryogenic properties of LH_2 tank, MDA and NASA investigated the compatibility of composite materials and liquid oxygen (LOx). The test results proved that composites could be used to fabricate Lox tanks for launch vehicles, and the results sufficiently convincing to plan on building and flying composite Lox tanks in X-33 and X-34 vehicles.

NASA and Lockheed Martin jointly developed the first composite cryogen tank for liquid oxygen storage. Lockheed Martin designed and manufactured a sub-scale LOX cryotank, while NASA tested it at its MFSC facility.

The composite tank is approximately 2.7 m in length and 1.2 m in diameter, and weighs less than 225 kg, the weight saving is 18% compared to metal tank of similar construction. The tank passed the initial testing of cyclic liquid oxygen loading. The tank withstood the thermal and pressure conditions similar to that on a space launch vehicle. The tank did not permeate nor crack after 52 cycles of fill-drain liquid oxygen test, and passed the demonstrated test, as shown in the figure below.

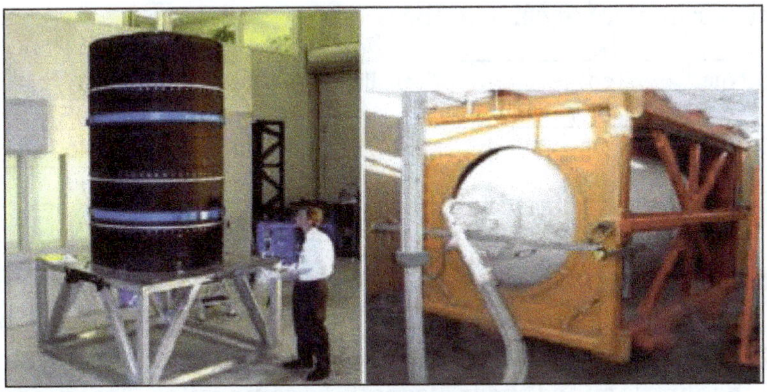

Composite lox tank of LM.

This tank marks a real advance in space technology. No approved standards for composite pressure vessels exist; there has not been enough information on them to write standards. So the technical data getting from this effort is very valuable.

Space X has designed, developed, and fabricated a composite LOx tank, which is a key component of interplanetary transport system (ITS). The tank is 12 m in diameter, and passed 2/3 exploration pressure test successfully in Nov 2016, see Figure. The carbon fiber used in the tank was provided by Toray. The 12-m diameter tank was the largest vessel ever produced.

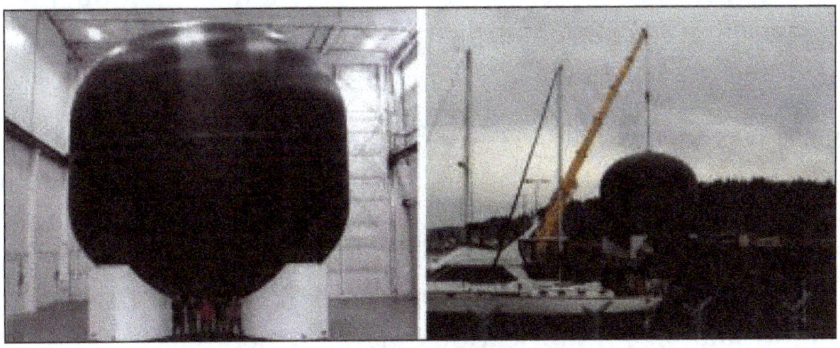

The 12-m diameter composite lox tank of space X.

Materials

Fiber-reinforced composites can be optimized and tailored with the right amount of fibers based on the directions and magnitudes of the stress state. Composite cryotanks can be the most efficient isotensoid structure. Yet the challenges remain with carbon fibers particularly for manufacturing large-scale cryotank.

IM7 which is manufactured by Hexcel, was used widely in cryogenic tank in NASA. IM600 which is made by Toho, was used in cryotank in Japan. T800H which is made in Toray, its property parameters are equivalent to IM7 and IM600. All of them are intermediate modulus fibers, high tensile strength, PAN-based fiber, their properties are basically equivalent.

Table: Property parameters of carbon fibers

	Fiber type	Tensile strength/MPa	Tensile modulus/GPa	Strain/%	Density/g/cm3	Filament diameter/μm
Toho	IM600	5790	285	2.0	1.80	5.0
Hercules	IM7	5300	275	1.8	1.77	5.2
Toray	T800H	5490	294	1.9	1.81	5.0

The carbon fiber with intermediate modulus, high tensile strength was used widely in composite cryotank manufacturing. The higher the modulus of carbon fiber, the higher the crystallinity of the fiber, this would lead to reduced surface active functional groups and decreased the interlaminar shear strength of composites. Low modulus carbon fiber is not good for the mechanical properties of composites. Therefore, intermediate modulus carbon fiber was employed for cryotank application.

Toughened resins were developed primarily for aircraft applications to improve the compression-after-impact strength (CAI) of composite structures. They are also preferred for liquid hydrogen tanks because of greater impermeability after thermo mechanical cycling. For example, 8552 epoxy resin, which was made by Hercules, 977-2 and 5320-1resins, which were made in ICI, were used widely in LH_2 cryotank fabrication.

Toughened resins appeared to be generally more resistant to ignition; therefore, toughened resins were used in LO2 tank fabrication. LOx compatibility is another issue that should be considered for Lox tank.

CF/977-2 prepreg was used widely in cryogenic applications because of excellent toughness and processing, for example, X-33 LH_2 tank and 10 m composite demonstrator tank made in CCTD phase I. CYCOM 977-2 is a 177°C curing toughed epoxy resin. It is formulated for autoclave or press molding. Unidirectional tape and woven fabric impregnated with CYCOM 977-2 resin will retain tack for at least 10 days and has a long mechanical out life suitable for fabrication of large structures.

The table below shows the properties of IM7/ 5320-1, IM7/977-2, and IM7/8552 laminates. Tensile property, compressive property, and shear property of them were basically equivalent, the CAI of IM7/ 5320-1 were the lowest. The advantage of 5320-1 is that it requires no autoclave curing.

Table: Properties parameters of IM7/5320-1and IM7/977-2 laminates.

	IM7/5320-1	IM7/977-2	IM7/8552
0° Tensile strength/MPa	2703	2690	2650
0° Tensile modulus/GPa	156	165	168
Poisson's ratio	0.34	—	—
90° Tensile strength/MPa	81	75	—

90°Tensile modulus/GPa	9.7	7.6	–
0°Compressive strength/MPa	1737	1580	1690
0°Compressive modulus/GPa	143	152	150
Short beam shear strength/MPa	119	112	128
CAI/MPa	176	262	234
Open-hole tensile strength/MPa	498	448	–
Open-hole compressive strength/MPa	386	310	–

References

- Cryogenics, Co-Di: scienceclarified.com, Retrieved 06 August, 2019
- Lee, Do W.; Jensen, Craig M. (2000). "Dry-Ice Bath Based on Ethylene Glycol Mixtures". J. Chem. Educ. 77: 629. doi:10.1021/ed077p629
- Pavel Urban; David Schmoranzer; Pavel Hanzelka; Katepalli R. Sreenivasan & Ladislav Skrbek (2013). "Anomalous heat transport and condensation in convection of cryogenic helium". Proceedings of the National Academy of Sciences. 110 (20): 8036–8039. Bibcode:2013PNAS..110.8036U. doi:10.1073/pnas.1303996110. PMC 3657834. PMID 23576759
- Stewart, MD; Koutroulakis, G; Kalechofsky, N; Mitrović, VF (2010). "A Reusable, Low-profile, Cryogenic Wire Seal". Cryogenics (Guildf). 50 (1): 50–51. doi:10.1016/j.cryogenics.2009.09.009. PMC 2805154. PMID 20161550
- The-application-of-carbon-fiber-composites-in-cryotank, solidification: intechopen.com, Retrieved 16 April, 2019
- A.T.A.M. de Waele, Basic operation of cryocoolers and related thermal machines, Review article, Journal of Low Temperature Physics, Vol.164, pp. 179-236, (2011), DOI: 10.1007/s10909-011-0373-x

Diverse Aspects of Cryobiology

CHAPTER 5

Cryobiology is a vast subject that has diverse aspects such as cryonics, cryoneurolysis, antifreeze protein, cryobank, cryochemistry, cryochemical synthesis, cryobiology and organ preservation, etc. The aim of this chapter is to explore these applications of cryobiology which are closely related to its progress.

CRYONICS

Cryonics is the practice of preserving human bodies in extremely cold temperatures with the hope of reviving them sometime in the future. The idea is that, if someone has "died" from a disease that is incurable today, he or she can be "frozen" and then revived in the future when a cure has been discovered. A person preserved this way is said to be in cryonic suspension.

To understand the technology behind cryonics, think about the news stories you've heard of people who have fallen into an icy lake and have been submerged for up to an hour in the frigid water before being rescued. The ones who survived did so because the icy water put their body into a sort of suspended animation, slowing down their metabolism and brain function to the point where they needed almost no oxygen.

Cryonics is a bit different from being resuscitated after falling into an icy lake, though. First of all, it's illegal to perform cryonic suspension on someone who is still alive. People who undergo this procedure must first be pronounced legally dead -- that is, their heart must have stopped beating. But if they're dead, how can they ever be revived? According to scientists who perform cryonics, "legally dead" is not the same as "totally dead." Total death, they say, is the point at which all brain function ceases. Legal death occurs when the heart has stopped beating, but some cellular brain function remains. Cryonics preserves the little cell function that remains so that, theoretically, the person can be resuscitated in the future.

ANTIFREEZE PROTEIN

Antifreeze proteins (AFPs) or ice structuring proteins (ISPs) refer to a class of polypeptides produced by certain animals, plants, fungi and bacteria that permit their survival in subzero environments. AFPs bind to small ice crystals to inhibit growth and recrystallization of ice that would otherwise be fatal. There is also increasing evidence that AFPs interact with mammalian cell membranes to protect them from cold damage. This work suggests the involvement of AFPs in cold acclimatization.

Non-colligative Properties

Unlike the widely used automotive antifreeze, ethylene glycol, AFPs do not lower freezing point in proportion to concentration. Rather, they work in a noncolligative manner. This phenomenon allows them to act as an antifreeze at concentrations 1/300th to 1/500th of those of other dissolved solutes. Their low concentration minimizes their effect on osmotic pressure. The unusual properties of AFPs are attributed to their selective affinity for specific crystalline ice forms and the resulting blockade of the ice-nucleation process.

Thermal Hysteresis

AFPs create a difference between the melting point and freezing point (busting temperature of AFP bound ice crystal) known as thermal hysteresis. The addition of AFPs at the interface between solid ice and liquid water inhibits the thermodynamically favored growth of the ice crystal. Ice growth is kinetically inhibited by the AFPs covering the water-accessible surfaces of ice.

Thermal hysteresis is easily measured in the lab with a nanolitre osmometer. Organisms differ in their values of thermal hysteresis. The maximum level of thermal hysteresis shown by fish AFP is approximately -3.5 °C (29.3 °F). However, insect antifreeze proteins are 10–30 times more active than fish proteins. This difference probably reflects the lower temperatures encountered by insects on land. In contrast, aquatic organisms are exposed only to -1 to -2 °C below freezing. During the extreme winter months, the spruce budworm resists freezing at temperatures approaching -30 °C. The Alaskan beetle Upis ceramboides can survive in a temperature of -60 °C by using antifreeze agents that are not proteins.

The rate of cooling can influence the thermal hysteresis value of AFPs. Rapid cooling can substantially decrease the nonequilibrium freezing point, and hence the thermal hysteresis value. Consequently, organisms cannot necessarily adapt to their subzero environment if the temperature drops abruptly.

Freeze Tolerance versus Freeze Avoidance

Species containing AFPs may be classified as:

- Freeze avoidant: These species are able to prevent their body fluids from freezing altogether. Generally, the AFP function may be overcome at extremely cold temperatures, leading to rapid ice growth and death.

- Freeze tolerant: These species are able to survive body fluid freezing. Some freeze tolerant species are thought to use AFPs as cryoprotectants to prevent the damage of freezing, but not freezing altogether. The exact mechanism is still unknown. However, it is thought AFPs may inhibit recrystallization and stabilize cell membranes to prevent damage by ice. They may work in conjunction with protein ice nucleators (PINs) to control the rate of ice propagation following freezing.

Diversity

There are many known nonhomologous types of AFPs.

Fish AFPs

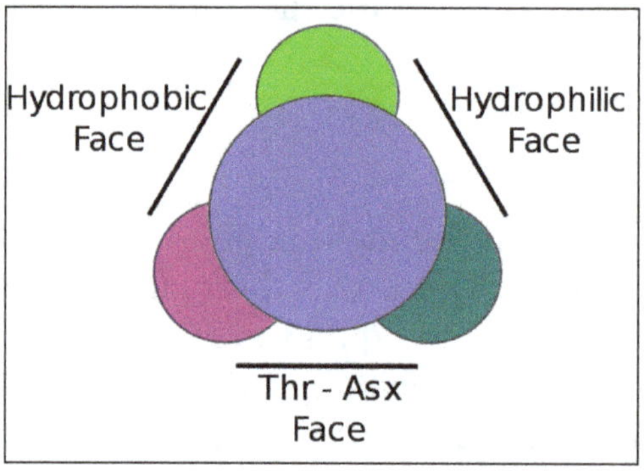

The three faces of Type I AFP.

Antifreeze glycoproteins or AFGPs are found in Antarctic notothenioids and northern cod. They are 2.6-3.3 kD. AFGPs evolved separately in notothenioids and northern cod. In notothenioids, the AFGP gene arose from an ancestral trypsinogen-like serine protease gene.

- Type I AFP is found in winter flounder, longhorn sculpin and shorthorn sculpin. It is the best documented AFP because it was the first to have its three-dimensional structure determined. Type I AFP consists of a single, long, amphipathic alpha helix, about 3.3-4.5 kD in size. There are three faces to the 3D structure: the hydrophobic, hydrophilic, and Thr-Asx face.
- Type I-hyp AFP (where hyp stands for hyperactive) are found in several righteye flounders. It is approximately 32 kD (two 17 kD dimeric molecules). The protein was isolated from the blood plasma of winter flounder. It is considerably better at depressing freezing temperature than most fish AFPs. The ability is partially derived from its many repeats of the Type I ice-binding site.
- Type II AFPs (e.g. P05140) are found in sea raven, smelt and herring. They are cysteine-rich globular proteins containing five disulfide bonds. Type II AFPs likely evolved from calcium dependent (c-type) lectins. Sea ravens, smelt, and herring are quite divergent lineages of teleost. If the AFP gene were present in

the most recent common ancestor of these lineages, it's peculiar that the gene is scattered throughout those lineages, present in some orders and absent in others. It has been suggested that lateral gene transfer could be attributed to this discrepancy, such that the smelt acquired the type II AFP gene from the herring.

- Type III AFPs are found in Antarctic eelpout. They exhibit similar overall hydrophobicity at ice binding surfaces to type I AFPs. They are approximately 6kD in size. Type III AFPs likely evolved from a sialic acid synthase (SAS) gene present in Antarctic eelpout. Through a gene duplication event, this gene—which has been shown to exhibit some ice-binding activity of its own—evolved into an effective AFP gene by loss of the N-terminal part.
- Type IV AFPs (P80961) are found in longhorn sculpins. They are alpha helical proteins rich in glutamate and glutamine. This protein is approximately 12KDa in size and consists of a 4-helix bundle. Its only posttranslational modification is a pyroglutamate residue, a cyclized glutamine residue at its N-terminus.

Plant AFPs

The classification of AFPs became more complicated when antifreeze proteins from plants were discovered. Plant AFPs are rather different from the other AFPs in the following aspects:

- They have much weaker thermal hysteresis activity when compared to other AFPs.
- Their physiological function is likely in inhibiting the recrystallization of ice rather than in the preventing ice formation.
- Most of them are evolved pathogenesis-related proteins, sometimes retaining antifungal properties.

Insect AFPs

There are two types of insect antifreeze proteins, Tenebrio and Dendroides AFPs which are both in different insect families. They are similar to one another, both being hyperactive (i.e. greater thermal hysteresis value) and consist of varying numbers of 12- or 13-mer repeats of approximately 8.3 to 12.5 kD. Throughout the length of the protein, at least every sixth residue is a cysteine. It has recently been discovered that even highly truncated insect AFP (e.g. peptide consisting of just two or three coils) may exhibit ice structuring and thermal hysteresis activities.

Tenebrio or Type V AFPs are found in beetles, whereas Dendroides or Choristoneura fumiferana AFPs are found in some Lepidoptera.

Sea Ice Organism AFPs

Many microorganisms living in sea ice possess AFPs that belong to a single family. The diatoms Fragilariopsis cylindrus and F. curta play a key role in polar sea ice communities, dominating the assemblages of both platelet layer and within pack ice. AFPs are widespread in these species, and the presence of AFP genes as a multigene family indicates the importance of this group for the genus Fragilariopsis. AFPs identified in F. cylindrus belong to an AFP family which is represented in different taxa and can be found in other organisms related to sea ice (Colwellia spp., Navicula glaciei, Chaetoceros neogracile and Stephos longipes and Leucosporidium antarcticum) and Antarctic inland ice bacteria (Flavobacteriaceae), as well as in cold-tolerant fungi (Typhula ishikariensis, Lentinula edodes and Flammulina populicola).

Several structures for sea ice AFPs have been solved. This family of proteins fold into a beta helix that form a flat ice-binding surface. Unlike the other AFPs, there is not a singular sequence motif for the ice-binding site.

Mechanisms of Action

AFPs are thought to inhibit growth by an adsorption–inhibition mechanism. They adsorb to nonbasal planes of ice, inhibiting thermodynamically favored ice growth. The presence of a flat, rigid surface in some AFPs seems to facilitate its interaction with ice via Van der Waals force surface complementarity.

Binding to Ice

Normally, ice crystals grown in solution only exhibit the basal (0001) and prism faces (1010), and appear as round and flat discs. However, it appears the presence of AFPs exposes other faces. It now appears the ice surface 2021 is the preferred binding surface, at least for AFP type I. Through studies on type I AFP, ice and AFP were initially thought to interact through hydrogen bonding. However, when parts of the protein thought to facilitate this hydrogen bonding were mutated, the hypothesized decrease in antifreeze activity was not observed. Recent data suggest hydrophobic interactions could be the main contributor. It is difficult to discern the exact mechanism of binding because of the complex water-ice interface. Currently, attempts to uncover the precise mechanism are being made through use of molecular modelling programs (molecular dynamics or the Monte Carlo method).

Binding Mechanism and Antifreeze Function

According to the structure and function study on the antifreeze protein from Pseudopleuronectes americanus, the antifreeze mechanism of the type-I AFP molecule was shown to be due to the binding to an ice nucleation structure in a zipper-like fashion through hydrogen bonding of the hydroxyl groups of its four Thr residues to the oxygens along the direction in ice lattice, subsequently stopping or retarding the growth of ice pyramidal planes so as to depress the freeze point.

The above mechanism can be used to elucidate the structure-function relationship of other antifreeze proteins with the following two common features:

- Recurrence of a Thr residue (or any other polar amino acid residue whose side-chain can form a hydrogen bond with water) in an 11-amino-acid period along the sequence concerned,
- A high percentage of an Ala residue component therein.

Commercial and Medical Applications

Numerous fields would be able to benefit from the protection of tissue damage by freezing. Businesses are currently investigating the use of these proteins in:

- Increasing freeze tolerance of crop plants and extending the harvest season in cooler climates,
- Improving farm fish production in cooler climates,
- Lengthening shelf life of frozen foods,
- Improving cryosurgery,
- Enhancing preservation of tissues for transplant or transfusion in medicine,
- Therapy for hypothermia,
- Human Cryopreservation (Cryonics).

Unilever has obtained UK approval to use a genetically modified yeast to produce antifreeze proteins from fish, for use in ice cream production.

HYPOTHERMIA

Hypothermia is defined as a body core temperature below 35.0 °C (95.0 °F) in humans. Symptoms depend on the temperature. In mild hypothermia there is shivering and mental confusion. In moderate hypothermia shivering stops and confusion increases. In severe hypothermia, there may be paradoxical undressing, in which a person removes their clothing, as well as an increased risk of the heart stopping.

Hypothermia has two main types of causes. It classically occurs from exposure to extreme cold. It may also occur from any condition that decreases heat production or increases heat loss. Commonly this includes alcohol intoxication but may also include low blood sugar, anorexia, and advanced age. Body temperature is usually maintained near a constant level of 36.5–37.5 °C (97.7–99.5 °F) through thermoregulation. Efforts to increase body temperature involve shivering, increased voluntary activity, and putting on warmer clothing. Hypothermia may be diagnosed based on either a person's symptoms in the presence of risk factors or by measuring a person's core temperature.

The treatment of mild hypothermia involves warm drinks, warm clothing, and physical activity. In those with moderate hypothermia, heating blankets and warmed intravenous fluids are recommended. People with moderate or severe hypothermia should be moved gently. In severe hypothermia, extracorporeal membrane oxygenation (ECMO) or cardiopulmonary bypass may be useful. In those without a pulse, cardiopulmonary resuscitation (CPR) is indicated along with the above measures. Rewarming is typically continued until a person's temperature is greater than 32 °C (90 °F). If there is no improvement at this point or the blood potassium level is greater than 12 mmol/liter at any time, resuscitation may be discontinued.

Hypothermia is the cause of at least 1,500 deaths a year in the United States. It is more common in older people and males. One of the lowest documented body temperatures from which someone with accidental hypothermia has survived is 13.0 °C (55.4 °F) in a near-drowning of a 7-year-old girl in Sweden. Survival after more than six hours of CPR has been described. For those for whom ECMO or bypass is used, survival is around 50%. Deaths due to hypothermia have played an important role in many wars.

Classification

Table: Hypothermia classification.

Hypothermia classification			
Swiss system	Symptoms	By degree	Temperature
Stage 1	Awake and shivering	Mild	32–35 °C (89.6–95.0 °F)
Stage 2	Drowsy and not shivering	Moderate	28–32 °C (82.4–89.6 °F)
Stage 3	Unconscious, not shivering	Severe	20–28 °C (68.0–82.4 °F)
Stage 4	No vital signs	Profound	<20 °C (68.0 °F)

Hypothermia is often defined as any body temperature below 35.0 °C (95.0 °F). With this method it is divided into degrees of severity based on the core temperature.

Another classification system, the Swiss staging system, divides hypothermia based on the presenting symptoms which is preferred when it is not possible to determine an accurate core temperature.

Other cold-related injuries that can be present either alone or in combination with hypothermia include:

- Chilblains: Condition caused by repeated exposure of skin to temperatures just above freezing. The cold causes damage to small blood vessels in the skin. This damage is permanent and the redness and itching will return with additional exposure. The redness and itching typically occurs on cheeks, ears, fingers, and toes.
- Frostbite: The freezing and destruction of tissue,
- Frostnip: A superficial cooling of tissues without cellular destruction,
- Trench foot or immersion foot: A condition caused by repetitive exposure to water at non-freezing temperatures.

The normal human body temperature is often stated as 36.5–37.5 °C (97.7–99.5 °F). Hyperthermia and fever, are defined as a temperature of greater than 37.5–38.3 °C (99.5–100.9 °F).

Signs and Symptoms

Signs and symptoms vary depending on the degree of hypothermia, and may be divided by the three stages of severity. Infants with hypothermia may feel cold when touched, with bright red skin and an unusual lack of energy.

Mild

Symptoms of mild hypothermia may be vague, with sympathetic nervous system excitation (shivering, high blood pressure, fast heart rate, fast respiratory rate, and contraction of blood vessels). These are all physiological responses to preserve heat. Increased urine production due to cold, mental confusion, and liver dysfunction may also be present. Hyperglycemia may be present, as glucose consumption by cells and insulin secretion both decrease, and tissue sensitivity to insulin may be blunted. Sympathetic activation also releases glucose from the liver. In many cases, however, especially in people with alcoholic intoxication, hypoglycemia appears to be a more common cause. Hypoglycemia is also found in many people with hypothermia, as hypothermia may be a result of hypoglycemia.

Moderate

As hypothermia progresses, symptoms include: mental status changes such as amnesia, confusion, slurred speech, decreased reflexes, and loss of fine motor skills.

Severe

As the temperature decreases, further physiological systems falter and heart rate, respiratory rate, and blood pressure all decrease. This results in an expected heart rate in the 30s at a temperature of 28 °C (82 °F).

There is often no shivering, cold, inflamed skin, hallucinations, lack of reflexes, fixed dilated pupils, low blood pressure, and pulmonary edema. Pulse and respiration rates decrease significantly, but fast heart rates (ventricular tachycardia, atrial fibrillation) can also occur. Atrial fibrillation is not typically a concern in and of itself.

Paradoxical Undressing

Twenty to fifty percent of hypothermia deaths are associated with paradoxical undressing. This typically occurs during moderate and severe hypothermia, as the person becomes disoriented, confused, and combative. They may begin discarding their clothing, which, in turn, increases the rate of heat loss.

Rescuers who are trained in mountain survival techniques are taught to expect this; however, people who die from hypothermia in urban environments are sometimes incorrectly assumed to have been subjected to sexual assault.

One explanation for the effect is a cold-induced malfunction of the hypothalamus, the part of the brain that regulates body temperature. Another explanation is that the muscles contracting peripheral blood vessels become exhausted (known as a loss of vasomotor tone) and relax, leading to a sudden surge of blood (and heat) to the extremities, causing the person to feel overheated.

Terminal Burrowing

An apparent self-protective behaviour, known as "terminal burrowing", or "hide-and-die syndrome", occurs in the final stages of hypothermia. The afflicted will enter small, enclosed spaces, such as underneath beds or behind wardrobes. It is often associated with paradoxical undressing. Researchers in Germany claim this is "obviously an autonomous process of the brain stem, which is triggered in the final state of hypothermia and produces a primitive and burrowing-like behavior of protection, as seen in hibernating animals". This happens mostly in cases where temperature drops slowly.

Causes

Hypothermia usually occurs from exposure to low temperatures, and is frequently complicated by alcohol consumption. Any condition that decreases heat production, increases heat loss, or impairs thermoregulation, however, may contribute. Thus, hypothermia risk factors include: substance abuse (including alcohol abuse), homelessness, any condition that affects judgment (such as hypoglycemia), the extremes of age, poor clothing, chronic medical conditions (such as hypothyroidism and sepsis), and living in a cold environment. Hypothermia occurs frequently in major trauma, and is also observed in severe cases of anorexia nervosa. Hypothermia is also associated with worse outcomes in people with sepsis. While most people with sepsis develop fevers (elevated body temperature), some develop hypothermia.

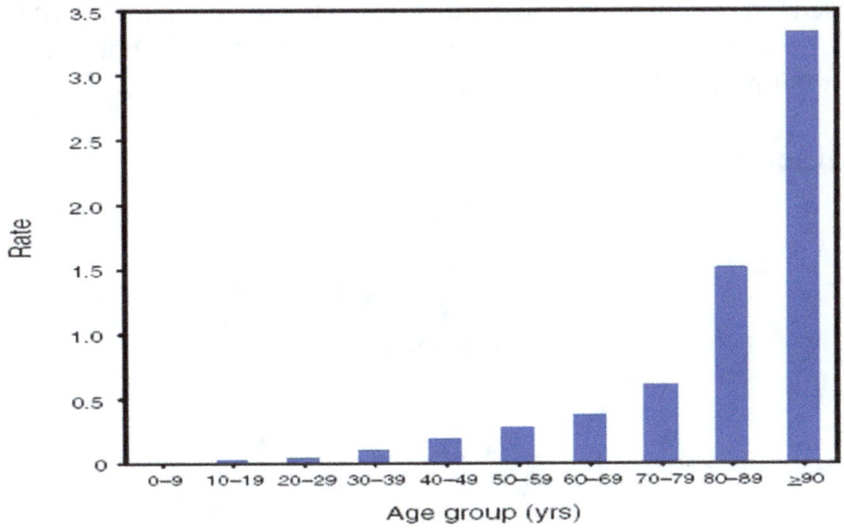

The rate of hypothermia is strongly related to age in the United States.

In urban areas, hypothermia frequently occurs with chronic cold exposure, such as in cases of homelessness, as well as with immersion accidents involving drugs, alcohol or mental illness. While studies have shown that homeless people are at risk of premature death from hypothermia, the true incidence of hypothermia-related deaths in this population is difficult to determine. In more rural environments, the incidence of hypothermia is higher among people with significant comorbidities and less able to move independently. With rising interest in wilderness exploration, and outdoor and water sports, the incidence of hypothermia secondary to accidental exposure may become more frequent in the general population.

Alcohol

Alcohol consumption increases the risk of hypothermia in two ways: vasodilation and temperature controlling systems in the brain. Vasodilation increases blood flow to the skin, resulting in heat being lost to the environment. This produces the effect of an individual feeling warm, when they are actually losing heat. Alcohol also affects the temperature-regulating system in the brain, decreasing the body's ability to shiver and use energy that would normally aid the body in generating heat. The overall effects of alcohol lead to a decrease in body temperature and a decreased ability to generate body heat in response to cold environments. Alcohol is a common risk factor for death due to hypothermia. Between 33% and 73% of hypothermia cases are complicated by alcohol.

Poverty

In the UK, 28,354 cases of hypothermia were treated in 2012–13 – an increase of 25% from the previous year. Some cases of hypothermia death, as well as other preventable deaths, happen because poor people cannot easily afford to keep warm. Rising fuel bills have increased the numbers who have difficulty paying for adequate heating in the UK. Some pensioners and disabled people are at risk because they do not work and cannot easily leave their homes. Better heat insulation can help.

Water Immersion

Two American marines participating in an immersion hypothermia exercise.

Hypothermia continues to be a major limitation to swimming or diving in cold water. The reduction in finger dexterity due to pain or numbness decreases general safety and work capacity, which consequently increases the risk of other injuries.

Other factors predisposing to immersion hypothermia include dehydration, inadequate rewarming between repetitive dives, starting a dive while wearing cold, wet dry suit undergarments, sweating with work, inadequate thermal insulation (for example, thin dry suit undergarment), and poor physical conditioning.

Heat is lost much more quickly in water than in air. Thus, water temperatures that would be quite reasonable as outdoor air temperatures can lead to hypothermia in survivors, although this is not usually the direct clinical cause of death for those who are not rescued. A water temperature of 10 °C (50 °F) can lead to death in as little as one hour, and water temperatures near freezing can cause death in as little as 15 minutes. A notable example of this occurred during the sinking of the Titanic, when most people who entered the −2 °C (28 °F) water died in 15–30 minutes.

The actual cause of death in cold water is usually the bodily reactions to heat loss and to freezing water, rather than hypothermia (loss of core temperature) itself. For example, plunged into freezing seas, around 20% of victims die within two minutes from cold shock (uncontrolled rapid breathing, and gasping, causing water inhalation, massive increase in blood pressure and cardiac strain leading to cardiac arrest, and panic); another 50% die within 15–30 minutes from cold incapacitation (inability to use or control limbs and hands for swimming or gripping, as the body "protectively" shuts down the peripheral muscles of the limbs to protect its core). Exhaustion and unconsciousness cause drowning, claiming the rest within a similar time.

Pathophysiology

Heat is primarily generated in muscle tissue, including the heart, and in the liver, while it is lost through the skin (90%) and lungs (10%). Heat production may be increased two- to four-fold through muscle contractions (i.e. exercise and shivering). The rate of heat loss is determined, as with any object, by convection, conduction, and radiation. The rates of these can be affected by body mass index, body surface area to volume ratios, clothing and other environmental conditions.

Many changes to physiology occur as body temperatures decrease. These occur in the cardiovascular system leading to the Osborn J wave and other dysrhythmias, decreased central nervous system electrical activity, cold diuresis, and non-cardiogenic pulmonary edema.

Research has shown that glomerular filtration rates (GFR) decrease as a result of hypothermia. In essence, hypothermia increases preglomerular vasoconstriction, thus decreasing both renal blood flow (RBF) and GFR.

Diagnosis

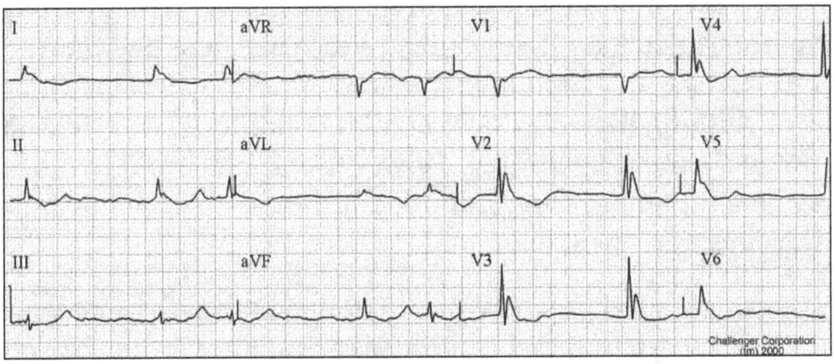

Atrial fibrillation and Osborn J waves in a person with hypothermia.
Note what could be mistaken for ST elevation.

Accurate determination of core temperature often requires a special low temperature thermometer, as most clinical thermometers do not measure accurately below 34.4 °C (93.9 °F). A low temperature thermometer can be placed in the rectum, esophagus or bladder. Esophageal measurements are the most accurate and are recommended once a person is intubated. Other methods of measurement such as in the mouth, under the arm, or using an infrared ear thermometer are often not accurate.

As a hypothermic person's heart rate may be very slow, prolonged feeling for a pulse could be required before detecting. In 2005, the American Heart Association recommended at least 30–45 seconds to verify the absence of a pulse before initiating CPR. Others recommend a 60-second check.

The classical ECG finding of hypothermia is the Osborn J wave. Also, ventricular fibrillation frequently occurs below 28 °C (82 °F) and asystole below 20 °C (68 °F). The Osborn J may look very similar to those of an acute ST elevation myocardial infarction. Thrombolysis as a reaction to the presence of Osborn J waves is not indicated, as it would only worsen the underlying coagulopathy caused by hypothermia.

Prevention

Appropriate clothing helps to prevent hypothermia. Synthetic and wool fabrics are superior to cotton as they provide better insulation when wet and dry. Some synthetic fabrics, such as polypropylene and polyester, are used in clothing designed to wick perspiration away from the body, such as liner socks and moisture-wicking undergarments. Clothing should be loose fitting, as tight clothing reduces the circulation of warm blood. In planning outdoor activity, prepare appropriately for possible cold weather. Those who drink alcohol before or during outdoor activity should ensure at least one sober person is present responsible for safety.

Covering the head is effective, but no more effective than covering any other part of the body. While common folklore says that people lose most of their heat through their heads, heat loss from the head is no more significant than that from other uncovered parts of the body. However, heat loss from the head is significant in infants, whose head is larger relative to the rest of the body than in adults. Several studies have shown that for uncovered infants, lined hats significantly reduce heat loss and thermal stress. Children have a larger surface area per unit mass, and other things being equal should have one more layer of clothing than adults in similar conditions, and the time they spend in cold environments should be limited. However children are often more active than adults, and may generate more heat. In both adults and children, overexertion causes sweating and thus increases heat loss.

Building a shelter can aid survival where there is danger of death from exposure. Shelters can be of many different types, metal can conduct heat away from the occupants and is sometimes best avoided. The shelter should not be too big so body warmth stays near the occupants. Good ventilation is essential especially if a fire will be lit in the shelter. Fires should be put out before the occupants sleep to prevent carbon monoxide poisoning. People caught in very cold, snowy conditions can build an igloo or snow cave to shelter.

The United States Coast Guard promotes using life vests to protect against hypothermia through the 50/50/50 rule: If someone is in 50 °F (10 °C) water for 50 minutes, he/she has a 50 percent better chance of survival if wearing a life jacket. A heat escape lessening position can be used to increase survival in cold water.

Babies should sleep at 16-20 °C (61-68 °F) and housebound people should be checked regularly to make sure the temperature of the home is at least 18 °C (64 °F).

Management

Aggressiveness of treatment is matched to the degree of hypothermia. Treatment ranges from noninvasive, passive external warming to active external rewarming, to active core rewarming. In severe cases resuscitation begins with simultaneous removal from the cold environment and management of the airway, breathing, and circulation. Rapid rewarming is then commenced. Moving the person as little and as gently as possible is recommended as aggressive handling may increase risks of a dysrhythmia.

Hypoglycemia is a frequent complication and needs to be tested for and treated. Intravenous thiamine and glucose is often recommended, as many causes of hypothermia are complicated by Wernicke's encephalopathy.

The UK National Health Service advises against putting a person in a hot bath, massaging their arms and legs, using a heating pad, or giving them alcohol. These measures can cause a rapid fall in blood pressure and potential cardiac arrest.

Rewarming

Rewarming can be done with a number of methods including passive external rewarming, active external rewarming, and active internal rewarming. Passive external rewarming involves the use of a person's own ability to generate heat by providing properly insulated dry clothing and moving to a warm environment. Passive external rewarming is recommended for those with mild hypothermia.

Active external rewarming involves applying warming devices externally, such as a heating blanket. These may function by warmed forced air (Bair Hugger is a commonly used device), chemical reactions, or electricity. In wilderness environments, hypothermia may be helped by placing hot water bottles in both armpits and in the groin. Active external rewarming is recommended for moderate hypothermia. Active core rewarming involves the use of intravenous warmed fluids, irrigation of body cavities with warmed fluids (the chest or abdomen), use of warm humidified inhaled air, or use of extracorporeal rewarming such as via a heart lung machine or extracorporeal membrane oxygenation (ECMO). Extracorporeal rewarming is the fastest method for those with severe hypothermia. When severe hypothermia has led to cardiac arrest, effective extracorporeal warming results in survival with normal mental function about 50% of the time. Chest irrigation is recommended if bypass or ECMO is not possible.

Rewarming shock (or rewarming collapse) is a sudden drop in blood pressure in combination with a low cardiac output which may occur during active treatment of a severely hypothermic person. There was a theoretical concern that external rewarming rather than internal rewarming may increase the risk. These concerns were partly believed to be due to afterdrop, a situation detected during laboratory experiments where there is a continued decrease in core temperature after rewarming has been started. Recent studies have not supported these concerns, and problems are not found with active external rewarming.

Fluids

For people who are alert and able to swallow, drinking warm sweetened liquids can help raise the temperature. Many recommend alcohol and caffeinated drinks be avoided. As most people are moderately dehydrated due to cold-induced diuresis, warmed intravenous fluids to a temperature of 38–45 °C (100–113 °F) are often recommended.

Cardiac Arrest

In those without signs of life, cardiopulmonary resuscitation (CPR) should be continued during active rewarming. For ventricular fibrillation or ventricular tachycardia, a single defibrillation should be attempted. However, people with severe hypothermia may not respond to pacing or defibrillation. It is not known if further defibrillation should be withheld until the core temperature reaches 30 °C (86 °F). In Europe, epinephrine is not recommended until the person's core temperature reaches 30 °C (86 °F), while the American Heart Association recommends up to three doses of epinephrine before a core temperature of 30 °C (86 °F) is reached. Once a temperature of 30 °C (86 °F) has been reached, normal ACLS protocols should be followed.

Prognosis

It is usually recommended not to declare a person dead until their body is warmed to a near normal body temperature of greater than 32 °C (90 °F), since extreme hypothermia can suppress heart and brain function. Exceptions include if there is an obvious fatal injuries or the chest is frozen so that it cannot be compressed. If a person was buried in an avalanche for more than 35 minutes and is found with a mouth packed full of snow without a pulse, stopping early may also be reasonable. This is also the case if a person's blood potassium is greater than 12 mmol/l.

Those who are stiff with pupils that do not move may survive if treated aggressively. Survival with good function also occasionally occurs even after the need for hours of CPR. Children who have near-drowning accidents in water near 0 °C (32 °F) can occasionally be revived, even over an hour after losing consciousness. The cold water lowers the metabolism, allowing the brain to withstand a much longer period of hypoxia. While survival is possible, mortality from severe or profound hypothermia remains high despite optimal treatment. Studies estimate mortality at between 38% and 75%.

In those who have hypothermia due to another underlying health problem, when death occurs it is frequently from that underlying health problem.

Other Animals

Many animals other than humans often induce hypothermia during hibernation or torpor.

Water bears (Tardigrade), microscopic multicellular organisms, can survive freezing at low temperatures by replacing most of their internal water with the sugar trehalose, preventing the crystallization that otherwise damages cell membranes.

CRYOBANK

A sperm bank, semen bank or cryobank is a facility or enterprise which purchases, stores and sells human semen. The semen is produced and sold by men who are known as sperm donors. The sperm is purchased by or for women for the purpose of achieving a pregnancy or pregnancies other than by a sexual partner. Sperm sold by a sperm donor is known as donor sperm. Sperm is introduced into the recipient woman by means of artificial insemination or by IVF and the process may also involve donated eggs or the use of a surrogate.

From a medical perspective, a pregnancy achieved using donor sperm is no different from a pregnancy achieved using partner sperm, and it is also no different from a pregnancy achieved by sexual intercourse. By using sperm from a donor rather than from the woman's partner, the process is a form of third party reproduction.

A sperm donor must generally meet specific requirements regarding age and medical history. In the United States, sperm banks are regulated as Human Cell and Tissue or Cell and Tissue Bank Product (HCT/Ps) establishments by the Food and Drug Administration. Many states also have regulations in addition to those imposed by the FDA.

In the European Union a sperm bank must have a license according to the EU Tissue Directive. In the United Kingdom, sperm banks are regulated by the Human Fertilisation and Embryology Authority.

Sperm banks provide the opportunity to have a baby to single women and coupled lesbians, and to heterosexual couples where the male is infertile. Where a sperm bank provides fertility services directly to a recipient woman, it may employ different methods of fertilization using donor sperm in order to optimise the chances of a pregnancy.

A sperm bank will also aim to provide donor sperm which is safe by the checking and screening donors and of their semen.

Some controversy stems from the fact that donors father children for others, often single women or coupled lesbians, but usually take no part in the upbringing of such children. The issue of sperm banks providing fertility services to single women and coupled lesbians so that they can have their own biological children by a donor is itself, often controversial. Donors may usually not have a say in who may be a recipient of their sperm.

Another controversy centers around the use of sperm of deceased men as pioneered by California Cryobank.

The increasing range of services which is available through sperm banks nevertheless enables more couples to have choices over the whole issue of reproduction. Women may choose an anonymous donor who will not be a part of family life, or they may choose known donors who may be contacted later in life by the donor children. Women may choose to use a surrogate to bear their children, using eggs provided by the woman and sperm from a donor. Sperm banks often provide services which enable a woman to have subsequent pregnancies by the same donor, but equally, women may choose to have children by a number of different donors. Sperm banks sometimes enable a woman to choose the sex of her child, enabling even greater control over the way families are planned. Sperm banks increasingly adopt a less formal approach to the provision of their services thereby enabling people to take a relaxed approach to their own individual requirements.

Men who donate their semen to a sperm bank do so with the intention that it will be used to enable women to have children whose partners have 'male factor' problems which prevent them from fathering children, or, more commonly, that they will enable women who have no male partner, such as single women and coupled lesbians, to have a child by them. Men who choose to donate semen through a sperm bank also have the security of knowing that they are helping such women or childless couples to have children in circumstances where they, as the biological father, will not have any legal or other responsibility for the children produced from their sperm. Whether a donor is anonymous or not, this factor is important in allowing sperm banks to recruit sperm donors and to use their sperm to produce whatever number of pregnancies from each donor as are permitted where they operate, or alternatively, whatever number they decide.

However, in many parts of the world sperm banks are not allowed to be established or to operate. Sperm banks do not provide a cure for infertility in that it is the sperm donor who reproduces himself, not a partner of the recipient woman. Most societies are built upon the family model and sperm banks may be seen as a threat to this, particularly where a sperm bank makes its services available to unmarried women.

Where sperm banks are allowed to operate they are often controlled by local legislation which is primarily intended to protect the unborn child, but which may also provide a compromise between the conflicting views which surround their operation. A particular example of this is the control which is often placed on the number of children which a single donor may father and which may be designed to protect against consanguinity. However, such legislation usually cannot prevent a sperm bank from supplying donor sperm outside the jurisdiction in which it operates, and neither can it prevent sperm donors from donating elsewhere during their lives. There is an acute shortage of sperm donors in many parts of the world and there is obvious pressure from many quarters for donor sperm from those willing and able to provide it to be made available as safely and as freely as possible.

Recruitment

The finding of a potential sperm donor and motivating him to actually donate sperm is typically called recruitment. A sperm bank can recruit donors by advertising, often in colleges and in local newspapers, and also on the internet.

A donor must be a fit healthy male, normally between 18 and 45 years of age, who is willing to undergo frequent and rigorous testing and who is willing to donate his sperm so that it can be used to impregnate women who are unrelated to and unknown by him. The donor must agree to relinquish all legal rights to all children which result from his donations. The donor must produce his sperm at the sperm bank thus enabling the identity of the donor, once proven, always to be ascertained, and also enabling fresh samples of sperm to be produced for immediate processing.Some sperm banks have been accused of heightism due to minimum height requirements.

Screening of Donors

A sperm donor must generally meet specific requirements regarding age and medical history.

Sperm banks typically screen potential donors for a range of diseases and disorders, including genetic diseases, chromosomal abnormalities and sexually transmitted infections that may be transmitted through sperm. The screening procedure generally also includes a quarantine period, in which the samples are frozen and stored for at least 6 months after which the donor will be re-tested for the STIs. This is to ensure no new infections have been acquired or have developed during the period of donation. Providing the result is negative, the sperm samples can be released from quarantine and used in treatments. Children conceived through sperm donation have a birth defect rate of almost a fifth compared with the general population.

A sperm bank takes a number of steps to ensure the health and quality of the sperm which it supplies and it will inform customers of the checks which it undertakes, providing relevant information about individual donors. A sperm bank will usually guarantee the quality and number of motile sperm available in a sample after thawing. They will try to select men as donors who are particularly fertile and whose sperm will survive the freezing and thawing process. Samples are often sold as containing a particular number of motile sperm per millilitre, and different types of sample may be sold by a sperm bank for differing types of use, e.g. |C| or |U|.

The sperm will be checked to ensure its fecundity and also to ensure that motile sperm will survive the freezing process. If a man is accepted onto the sperm bank's program as a sperm donor, his sperm will be constantly monitored, the donor will be regularly checked for infectious diseases, and samples of his blood will be taken at regular intervals. A sperm bank may provide a donor with dietary supplements containing herbal or mineral substances such as maca, zinc, vitamin E and arginine which are designed to improve the quality and quantity of the donor's semen, as well as reducing the refractory time (i.e. the time between viable ejaculations). All sperm is frozen in straws or vials and stored for as long as the sperm donor may and can maintain it.

Donors are subject to tests for infectious diseases such as human immunoviruses HIV (HIV-1 and HIV-2), human T-cell lymphotropic viruses (HTLV-1 and HTLV-2), syphilis, chlamydia, gonorrhea, Hepatitis B virus, Hepatitis C virus, cytomegalovirus (CMV), Trypanosoma cruzi and Malaria as well as hereditary diseases such as cystic fibrosis, Sickle cell anemia, Familial Mediterranean fever, Gaucher's disease, Thalassaemia, Tay–Sachs disease, Canavan's disease, Familial dysautonomia, Congenital adrenal hyperplasia Carnitine transporter deficiency and Karyotyping 46XY. Karyotyping is not a requirement in either EU or the US but some sperm banks choose to test donors as an extra service to the customer.

A sperm donor may also be required to produce his medical records and those of his family, often for several generations. A sperm sample is usually tested micro-biologically at the sperm bank before it is prepared for freezing and subsequent use. A sperm donor's blood group may also be registered to ensure compatibility with the recipient.

Some sperm banks may disallow sexually active gay men from donating sperm due to the population's increased risk of HIV and hepatitis B. Modern sperm banks have also been known to screen out potential donors based on genetic conditions and family medical history.

Donor Payment

The majority of sperm donors who donate their sperm through a sperm bank receive some kind of payment, although this is rarely a significant amount. A review including 29 studies from 9 countries came to the result that the amount of money actual donors received for their donation varied from $10 to €70 per donation or sample. The payments vary from the situation in the United Kingdom where donors are only entitled to their expenses in connection with the donation, to the situation with some US sperm banks where a donor receives a set fee for each donation plus an additional amount for each vial stored. At one prominent California sperm bank for example, TSBC, donors receive roughly $50 for each donation (ejaculation) which has acceptable motility/survival rates both at donation and at a test-thaw a couple of days later. Because of the requirement for the two-day celibacy period before donation, and geographical factors which usually require the donor to travel, it is not a viable way to earn a significant income—and is far less lucrative than selling human eggs. Some private donors may seek remuneration although others donate for altruistic reasons. According to the EU Tissue Directive donors in EU may only receive compensation, which is strictly limited to making good the expenses and inconveniences related to the donation. A survey among sperm donors in Cryos International Sperm bank showed that altruistic as well as financial motives were the main factors for becoming a donor. However, when the compensation was increased 100% in 2004 (to DKK 500) it had no significant impact on either the numbers of new donor candidates coming in or the frequency of donations from the existing donors. When the compensation was reduced to the previous level (DKK 250) again one year later in 2005 there was no effect either. This led to the assumption that altruism is the main motive and that financial compensation is secondary.

Collection

A sperm donor will usually be required to enter into a contract with a sperm bank to supply his semen, typically for a period of six to twenty-four months depending on the number of pregnancies which the sperm bank intends to produce from the donor. Where local regulations or the sperm bank's own rules limit the number of pregnancies which a single donor can achieve, his donations will be limited for this reason. In the United Kingdom, for example, where a donor is not permitted to father more than ten families, a sperm bank will generally need a maximum of 100 straws prepared for IUI insemination, so that a man will generally not donate for more than twelve months, unless the sperm bank exports or exchanges sperm with sperm banks outside the UK.

However, not all donors complete the intended program of donations. If a sperm bank has access to world markets e.g. by direct sales, or sales to clinics outside their own jurisdiction, a man may donate for a longer period than two years, as the risk of consanguinity is reduced (although local laws vary widely). Some sperm banks with access to world markets impose their own rules on the number of pregnancies which can be achieved in a given regional area or a state or country, and these sperm banks may permit donors to donate for four or five years, or even longer. Faced with a growing demand for donor sperm, sperm banks may try to maximise the use of a donor whilst still reducing the risk of consanguinity.

The contract may also specify the place and hours for donation, a requirement to notify the sperm bank in the case of acquiring a sexual infection, and the requirement not to have intercourse or to masturbate for a period of usually 2–3 days before making a donation.

A sperm donor generally produces and collects sperm at a sperm bank or clinic by masturbation in a private room or cabin, known as a 'men's production room' (UK), 'donor cabin' (DK) or a masturbatorium (USA). Many of these facilities contain pornography such as videos/DVD, magazines, and/or photographs which may assist the donor in becoming aroused in order to facilitate production of the ejaculate, also known as the "semen sample". In some circumstances, it may also be possible for semen from donors to be collected during sexual intercourse with the use of a collection condom.

Processing Sperm

Sperm banks and clinics usually 'wash' the sperm sample to extract sperm from the rest of the material in the semen. A cryoprotectant semen extender is added if the sperm is to be placed in frozen storage. One sample can produce 1–20 vials or straws, depending on the quantity of the ejaculate and whether the sample is 'washed' or 'unwashed'. 'Unwashed' samples are used for intracervical insemination (ICI) treatments, and 'washed' samples are used in intrauterine insemination (IUI) and for in-vitro fertilization (IVF) procedures.

Storage

The sperm is stored in small vials or straws holding between 0.4 and 1.0 ml of sperm and cryogenically preserved in liquid nitrogen tanks. It has been proposed that there should be an upper limit on how long frozen sperm can be stored; however, a baby has been conceived in the United Kingdom using sperm frozen for 21 years and andrology experts believe sperm can be frozen indefinitely. The UK government places an upper limit for storage of 55 years.

Before freezing, sperm may be prepared (washed or left unwashed) so that it can be used for intracervical insemination (ICI), intrauterine insemination (IUI) or for in-vitro fertilization (IVF) or assisted reproduction technologies (ART).

Following the necessary quarantine period, which is usually 6 months, a sample will be thawed and used to artificially inseminate a woman or used for another assisted reproduction technologies (ART) treatment.

Services

Use

Subject to any regulations restricting who can obtain donor sperm, donor sperm is available to all women who, for whatever reason, wishes to have a child. These regulations vary significantly between jurisdictions, and some countries do not have any regulations. When a woman finds that she is barred from receiving donor sperm within her jurisdiction, she may travel to another jurisdiction to obtain sperm. Regulations change from time to time. In most jurisdictions, donor sperm is available to a woman if her partner is infertile or where he has a genetic disorder. However, the categories of women who may obtain donor sperm is expanding, with its availability to single women and to lesbian couples becoming more common, and some sperm banks supply fertility centres which specialise in the treatment of such women.

Frozen vials of donor sperm may be shipped by the sperm bank to a recipient's home for self-insemination, or they may be shipped to a fertility clinic or physician for use in fertility treatments. The sperm bank will rely on the recipient woman or medical practitioner to report the outcome of any use of the sperm to the sperm bank. This enables to sperm bank to adhere to any national limits of pregnancy numbers. The sperm bank may also impose its own worldwide limit on numbers.

Men may also store their own sperm at a sperm bank for future use particularly where they anticipate traveling to a war zone or having to undergo chemotherapy which might damage the testes.

Sperm from a sperm donor may also be used in surrogacy arrangements and for creating embryos for embryo donation. Donor sperm may be supplied by the sperm bank directly to the recipient to enable a woman to perform her own artificial insemination which can be carried out using a needleless syringe or a cervical cap conception device. The cervical cap conception device allows the donor semen to be held in place close to the cervix for between six and eight hours to allow fertilization to take place. Alternatively, donor sperm can be supplied by a sperm bank through a registered medical practitioner who will perform an appropriate method of insemination or IVF treatment using the donor sperm in order for the woman to become pregnant.

Choosing Donors

Information about Donor

In the United States, sperm banks maintain lists or catalogs of donors which provide basic information about the donor such as racial origin, skin color, height, weight, colour of eyes, and blood group. Some of these catalogs are available for browsing on the Internet, while others are made available to patients only when they apply to a sperm bank for treatment. Some sperm banks make additional information about each donor available for an additional fee, and others make additional basic information known to children produced from donors when those children reach the age of 18. Some clinics offer "exclusive donors" whose sperm is used to produce pregnancies for only one recipient woman. How accurate this is, or can be, is not known, and neither is it known whether the information produced by sperm banks, or by the donors themselves, is true. Many sperm banks will, however, carry out whatever checks they can to verify the information they request, such as checking the identity of the donor and contacting his own doctor to verify medical details.

In the United Kingdom, most donors are anonymous at the point of donation and recipients can see only non-identifying information about their donor (height, weight, ethnicity etc.). Donors need to provide identifying information to the clinic and clinics will usually ask the donor's doctor to confirm any medical details they have been given. Donors are asked to provide a pen portrait of themselves which is held by the HFEA and can be obtained by the adult conceived from the donation at the age of 18, along with identifying information such as the donor's name and last known address. Known donation is permitted and it is not uncommon for family or friends to donate to a recipient couple.

Qualities that potential recipients typically prefer in donors include the donors being tall, college educated, and with a consistently high sperm count. A review came to the result that 68% of donors had given information to the clinical staff regarding physical characteristics and education but only 16% had provided additional information such as hereditary aptitudes and temperament or character.

Recipient's Selection of Donors

Sperm banks make information available about the sperm donors whose donations they hold to enable customers to select the donor whose sperm they wish to use. This information is often available by way of an online catalog. A sperm bank will also usually have facilities to help customers to make their choice and they will be able to advise on the suitability of donors for individual donors and their partners.

Where the recipient woman has a partner she may prefer to use sperm from a donor whose physical features are similar to those of her partner. In some cases, the choice of a donor with the correct blood group will be paramount, with particular considerations for the protection of recipients with negative blood groups. If a surrogate is to be used, such as where the customer is not intending to carry the child, considerations about her blood group etc. will also need to be taken into account.

Information made available by a sperm bank will usually include the race, height, weight, blood group, health and eye colour of the donor. Sometimes information about his age, family history and educational achievements will also be given. Some sperm banks make a 'personal profile' of a donor available and occasionally more information may be purchased about a donor, either in the form of a DVD or in written form. Catalogs usually state whether samples supplied in respect of a particular donor have already given rise to pregnancies, but this is not necessarily a guide to the fecundity of the sperm since a donor may not have been in the program long enough for any pregnancies to have been recorded.

If a woman intends to have more than one child, she may wish to have the additional child or children by the same donor. Sperm banks will usually advise whether sufficient stocks of sperm are available from a particular donor for subsequent pregnancies, and they normally have facilities available so that the woman may purchase and store additional vials from that donor on payment of an appropriate fee. These will be stored until required for subsequent pregnancies or they may be onsold if they become surplus to the woman's requirements.

The catalogue will also state whether samples of sperm are available for ICI, IUI, or IVF use.

Sex Selection

Some sperm banks enable recipients to choose the sex of their child, through methods of sperm sorting. Although the methods used do not guarantee 100% success, the chances of being able to select the gender of a child are held to be considerably increased.

One of the processes used is the 'swim up' method, whereby a sperm extender is added to the donor's freshly ejaculated sperm and the test-tube is left to settle. After about half-an-hour, the lighter sperm, containing the male chromosome pair (XY), will have swum to the top, leaving the heavier sperm, containing the female chromosome pair (XX), at the bottom, thus allowing selection and storage according to sex.

The alternative process is the Percoll Method which is similar to the 'swim up' method but involves additionally the centrifuging of the sperm in a similar way to the washing of samples produced for IUI inseminations, or for IVF purposes.

Sex selection is not permitted in a number of countries, including the UK.

Other Uses

There is a market for vials of processed sperm and for various reasons a sperm bank may sell-on stocks of vials which it holds (known as 'onselling'). The costs of screening of donors and of storage of frozen donor sperm vials are not insignificant and in practice most sperm banks will try to dispose of all samples from an individual donor. The onselling of sperm therefore enables a sperm bank to maximize the sale and disposal of sperm samples which it has processed. The reasons for onselling may also be where part of, or even the main business of, a particular sperm bank is to process and store sperm rather than to use it in fertility treatments, or where a sperm bank is able to collect and store more sperm than it can use within nationally set limits. In the latter case a sperm bank may onsell sperm from a particular donor for use in another jurisdiction after the number of pregnancies achieved from that donor has reached its national maximum.

Sperm banks may supply other sperm banks or a fertility clinic with donor sperm to be used for achieving pregnancies.

Sperm banks may also supply sperm for research or educational purposes.

CRYOCHEMISTRY

Cryochemistry is the study of chemical interactions at temperatures below −150 °C (−238 °F; 123 K). It overlaps with many other sciences, including chemistry, cryobiology, condensed matter physics, and even astrochemistry.

Cryochemistry has been a topic of interest since liquid nitrogen, which freezes at −210°C, became commonly available. Cryogenic-temperature chemical interactions are an important mechanism for studying the detailed pathways of chemical reactions by reducing the confusion introduced by thermal fluctuations. Cryochemistry forms the foundation for cryobiology, which uses slowed or stopped biological processes for medical and research purposes.

Low Temperature Behaviours

When cooled below the Lambda point (2.17 K or −270.98 °C at 1 atm), liquid helium exhibits properties of superfluidity.

As a material cools, the relative motion of its component molecules/atoms decreases - its temperature decreases. Cooling can continue until all motion ceases, and its kinetic energy, or energy of motion, disappears. This condition is known as absolute zero and it forms the basis for the Kelvin temperature scale, which measures the temperature above absolute zero. Zero degrees Celsius (°C) coincides with 273 Kelvin.

At absolute zero most elements become a solid, but not all behave as predictably as this; for instance, helium becomes a highly unusual liquid. The chemistry between substances, however, does not disappear, even near absolute zero temperatures, since separated molecules/atom can always combine to lower their total energy. Almost every molecule or element will show different properties at different temperatures; if cold enough, some functions are lost entirely. Cryogenic chemistry can lead to very different results compared with standard chemistry, and new chemical routes to substances may be available at cryogenic temperatures, such as the formation of argon fluorohydride, which is only a stable compound at or below 17 K (−256.1 °C).

Methods of Cooling

One method that used to cool molecules to temperatures near absolute zero is laser cooling. In the Doppler cooling process, lasers are used to remove energy from electrons of a given molecule to slow or cool the molecule down. This method has applications in quantum mechanics and is related to particle traps and the Bose–Einstein condensate. All of these methods use a "trap" consisting of lasers pointed at opposite equatorial angles on a specific point in space. The wavelengths from the laser beams eventually hit the gaseous atoms and their outer spinning electrons. This clash of wavelengths decreases the kinetic energy state fraction by fraction to slow or cool the molecules down. Laser cooling has also been used to help improve atomic clocks and atom optics. Ultracold studies are not usually focused on chemical interactions, but rather on fundamental chemical properties.

Because of the extremely low temperatures, diagnosing the chemical status is a major issue when studying low temperature physics and chemistry. The primary techniques in use today are optical - many types of spectroscopy are available, but these require special apparatus with vacuum windows that provide room temperature access to cryogenic processes.

CRYOCHEMICAL SYNTHESIS

Cryochemical synthesis also called cryochemical processing or cryoprocessing refers to the set of methods to synthesise substances and materials based on low-temperature chemical processes.

The most widespread of such methods are cryochemical synthesis techniques with the use of aqueous solutions. In this case, solutions, suspensions or products of chemical co-precipitation of initial solution components containing cathions of the synthesised material in a stoichiometric proportion are rapidly frozen and freeze dried in vacuum, after which thermal decomposition follows. The products of this synthesis are generally oxide powders with crystallite sizes of 40-300 nm whose degree of agglomeration depends on the choice of substances frozen in the process (solution/suspension/sludge). It is also possible to remove ice by low-temperature extraction in polar organic solvents (cryoextraction) and coprecipitation of frozen components in low temperature precipitant solution (cryoprecipitation).

When applying catalysts on porous supports it is advisable to use the cryoimpregnation method, where substrates soaked with a solution of catalytically active components undergo rapid freezing and freeze-drying. The use of low temperatures can significantly increase the homogeneity of the micro component distribution on the substrate's surface. Cryochemical methods also include low-temperature mechanochemical techniques, in particular, cryomilling.

Deposition of metal vapors on very cold surfaces in the presence of inert gases or organic substances is an efficient method to produce nanoparticles of metals ranging from 1 to 10 nm in size.

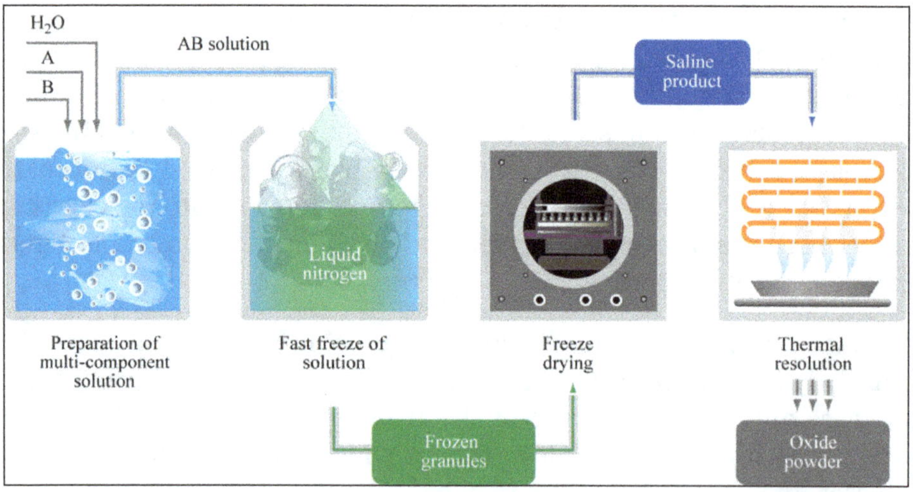

Basic processes of cryochemical synthesis involving frozen aqueous solutions.

CRYOBIOLOGY AND ORGAN PRESERVATION

Despite the continued need for organ availability, survivorship after transplantation is quite high, with ≥75 % of those receiving kidney, heart, and liver transplants living more than 5 years after the transplantation surgery.

There are a number of logistical issues associated with organ transplantation, all of which contribute to the challenge of improving the outcomes for those waiting for transplants.

It is unlikely that perfect equity in regards to distribution of available organs will ever be achieved, despite the best efforts of organizations in this particular area. One significant reason for this is the finite time for which organs remain viable after the onset of ischemia, often measured in hours, not days. As a result, in order for the best outcomes to be achieved, the transplant must take place as quickly as possible after organ procurement. For example, in one study, it was reported that twice the number of organs showed signs of change when transported to a different facility from where the procurement took place compared to those having been transplanted at the same facility.

The heart and lung have the shortest viable lifespan during ischemic cold storage, with viability dropping off after ~4 and 6–8 h, respectively. Kidneys, on the other hand, tolerate cold ischemia better, and it is considered acceptable to transplant kidneys within 18 h of procurement. The liver and pancreas fall between these two values, with ideal transplantation times being within 12 h of procurement. For organs having been donated after circulatory death of the donor individual, these numbers are reduced.

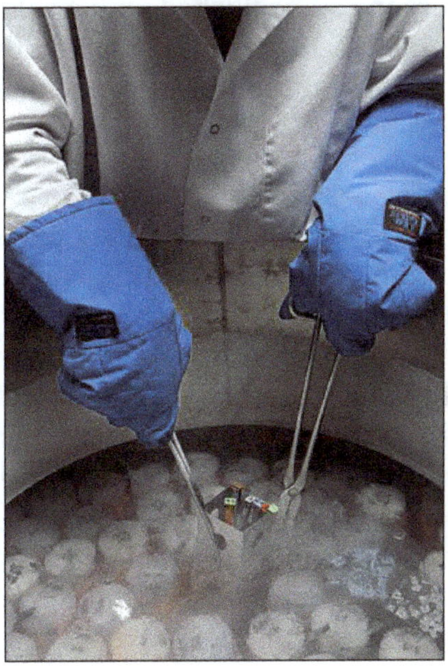

Significant advances in preservation have occurred over the previous decades, allowing extension of the viable time frame between organ procurement and transplantation. This has been due to the development of cold preservation solutions and means of their application.

Transplantation and Organ Damage

Ischemia and Hypothermic Injury

Since the early days of organ transplantation, people have studied the consequences of temporary and long-term storage of organs to facilitate the procedure. Over the years, this has resulted in many advances in our understanding of the pathophysiology of what has become to be known as ischemia/reperfusion (I/R) injury. The greater understanding of these consequences and their mechanisms create promising possibilities for reducing sources of injury during organ preservation and transplantation, as has opened opportunities for other strategies as well, such as tissue preconditioning to tolerate I/R and methods to repair injured tissues directly after the fact. Organs can be injured beyond repair in response to both ischemia and, paradoxically, restoration of blood flow. I/R injury is especially relevant to complex tissue allografts (CTAs) such as vascularized tissues or organs. Ischemia describes deficient blood supply and has been used to describe the process since the early nineteenth century. Since first understanding the effect it had on organs, researchers have endeavored to understand its underlying mechanisms.

Organ ischemia can be divided into two phases. The first is warm ischemia, in which an organ may be exposed to relatively high temperatures immediately following disruption of circulation until the organ can be flushed with a cold, hypothermic preservation solution. The second phase, cold ischemia, occurs when the organ is maintained in a hypothermic state prior to transplantation into the recipient. In multiorgan recovery, the organs are typically cooled before they are removed.

Cooling is indeed the best defense against ischemic injury; however, lowering the temperature of a biological system wreaks havoc on many fundamentally important structural and functional aspects. For instance, the cellular membrane plays a structural role and provides an active interface with the extracellular environment. Proteins responsible for cell signaling, ion regulation, and other important processes are integrally linked to the cellular membrane and essential to cellular function. The ability of the membrane to protect the cell from ionic and molecular permeability (including water permeability) depends on the integrity of the lipid bilayer and on its ability to maintain tight control of these parameters. Low temperature can result in a membrane phase transition of lipids and can drastically affect cellular stability. Low temperature can also impair membrane-bound enzymes crucial for maintaining cellular homeostasis. Membrane changes caused by such phase transitions are known to cause increases in membrane permeability. If normally non-permeable species enter the cell during the phase transition, they become trapped upon return to ambient or body temperature. The result of this is cell swelling when water rushes into the intracellular space to balance the chemical disequilibrium, which causes cell swelling and potentially lysis. This can be mitigated by adding benign non-permeating solutes to organ preservation solutions.

During ischemia, excessive production of hydrogen ions occurs as the cell is forced into anaerobic metabolism. This results in a lack of adenosine triphosphate (ATP) and a resulting impairment of the sodium-potassium adenosine triphosphatase (Na-K ATPase) pump. Impairment of this pump, which functions to maintain the ionic composition of the cell, causes potassium to move out of the intracellular space through osmotic diffusion, and sodium, which is normally found in relatively low concentrations inside the cell, to diffuse in. This change in intracellular ionic composition can cause swelling and potentially cell lysis. To mitigate this effect of hypothermic ischemia, use of preservation solutions with electrolyte compositions similar to the intracellular composition (e.g., high potassium and low sodium) is often employed and allows the cell to rest in a state of inactive homeostasis.

Hydrogen-ion production increases in ischemic organs causing intracellular pH to decrease. In the liver especially, where the switch from aerobic to anaerobic glycolysis occurs during ischemia and results in increasing production of lactic acid, this type of injury is particularly problematic. Additionally, the membrane permeability of calcium is increased during ischemia. As this occurs, influx of calcium can lead to intracellular acidosis. Calmodulin, which senses these high calcium levels, relays signals to various calcium-sensitive enzymes which, in turn, cause upregulation of phospholipases. This can lead to mitochondrial membrane permeability and subsequent dysfunction. Increased cellular calcium concentrations also initiate myofibrillar contraction of the vascular smooth muscle, causing vasospasm and subsequent ischemic damage. Endothelin, a 21-amino acid peptide produced by the vascular endothelium, can exacerbate ischemic vasoconstriction and can therefore worsen ventricular, coronary, and other organ function upon reperfusion. Endothelin A and B receptor antagonists have been evaluated successfully to improve results after solid organ transplantation.

Hypothermia also decreases the metabolic rate and slows degradation of cellular components, but even at low (above freezing) temperatures, metabolism is not completely suppressed. Aerobic cells function based on a combination of the anaerobic enzymatic breakdown of glucose (glycolysis) and aerobic cellular respiration. Cooling from 37 to 0 °C reduces cellular metabolism 12-fold. Because of this, even though metabolism and utilization of cellular energy stores are slowed, ATP and adenosine diphosphate (ADP), the major sources of cellular metabolic energy, are gradually depleted during hypothermia. This depletion is a result of the residual energy requirements of the cell at low temperature still exceeding the amount of ATP able to be produced by the cell. During organ ischemia, anaerobic gycolysis results in lactate production and impairment of mitochondrial respiration. This ischemic mitochondrial dysfunction is responsible for most of the damage due to cellular energy disruption, and overall reduction of mitochndrial enzyme activity and/or mitochondrial alteration has been associated with apoptotic pathways. Recent studies evaluating mitochondrial-targeted antioxidants show promise in mitigating this type of damage.

Reperfusion Injury

Paradoxically, the major injures to transplanted organs occur during reperfusion as opposed to ischemia. In addition to frank injury, some of the events that occur during reperfusion may result in enhanced immunogenicity of the graft. These findings have led to many advances in organ preservation aimed at preventing reperfusion injury. Oxygen free radicals generated during reperfusion are arguably the main cause of the reperfusion injury, acting through various mechanisms. While an organ experiences ischemia, increased calcium levels can activate specific intracellular enzymes that convert xanthine dehydrogenase to xanthine oxidase. Both these enzymes catabolize hypoxanthine and xanthine to uric acid. In the case of xanthine oxidase, molecular oxygen is used as an electron acceptor which results in formation of superoxide. Superoxide rapidly forms hydrogen peroxide, which is a strong oxidant that can cause cell injury or death by oxidizing lipid membranes and cellular proteins directly, and produces many other reactive oxygen species, including hydroxyl radical and singlet oxygen. Oxygen free radicals form very quickly, and the damaging effects on the organ begin almost immediately upon reperfusion.

Additionally, when an organ is exposed to ischemia, oxygen levels become too low for xanthine oxidase to metabolize xanthine and hypoxanthine, causing a buildup in the intracellular concentration of these metabolites. Upon reperfusion when oxygen is once again available, cellular systems proceed and the backlog of metabolism results in a sudden production of reactive oxygen intermediates. This overwhelms the cellular pathways responsible for scavenging reactive oxygen species and results in cellular injury. Inhibitors of xanthine oxidase such as allopurinol have been shown to protect against this type of ischemic injury when applied prior to the insult.

Reactive oxygen species can also result in lipid peroxidation. During lipid peroxidation, oxygen free radicals interact with polyunsaturated fatty acids in the cellular membrane resulting in a chain reaction that may ultimately destroy cellular integrity and thus kill the cell. Glutathione, an endogenous free radical scavenger, can reduce lipid peroxidation and is used along with other similar agents in organ preservation solutions to mitigate reperfusion injury.

Oxygen free radicals can also directly activate phospholipase A201 which can in turn initiate production of prostaglandins such as leukotriene B4. This specific prostaglandin is a chemokine that causes leukocytes to adhere to the vascular endothelium, which can result in degranulation causing proteolytic damage and/or blocking of microcirculation.

In addition to reactive oxygen species, cytokines and nitric oxide also play a significant role in reperfusion injury. Cytokines are intercellular messengers that are produced both during normal as well as pathophysiologic states. Ischemia/reperfusion injury results in a dramatic increase of several such cytokines, including tumor necrosis factor (TNF)-alpha, interferon-gamma, interleukin-1, and interleukin-8. Increased production of these cytokines results in upregulation of adhesion molecules that can cause leukocyte adherence and platelet plugging following revascularization, which can result in graft failure and/or rejection. Likewise, nitric oxide (NO) has been correlated with acute rejection. An extremely reactive autocoid, NO is generated by nitric oxide synthase from L-arginine. Reports indicate that NO production is induced by inflammatory cytokines, such as TNF-alpha, interferon-gamma, and interleukin-1.

Challenges with Long-term Preservation of Organs

While the time frames for which organs can be preserved outside of the normal physiological environment have been extended it is likely that the only means to optimize the utility of organs for donation would be to preserve them in a manner that eliminates deterioration on any meaningful time scale. The use of cooling to slow down metabolism and reduce the injury associated with the lack of adequate energy supplies can be extrapolated in such a manner that all forms of degradation can be stopped. Theoretically, this could be achieved with the use of cryopreservation, as is commonly done with isolated cells and small tissue fragments on a routine basis.

A number of challenges exist with the cryopreservation of large pieces of tissue on the scale of whole organs, some of which are similar to the challenges with cryopreserving any biological cell and others that scale with the size of the material.

Since the publication in 1949 of the serendipitous discovery of the cryoprotective effects of glycerol on spermatozoa, it has been understood that, with very few exceptions, the use of intracellular cryoprotective compounds is necessary to preserve cells from the injurious effects of freezing. While we still have an incomplete understanding of the extent of their modes of action, we are aware of a few means by which they help to improve the viability of cells after their use.

One of the earliest hypotheses tested was that they have a colligative effect and reduce the concentration of salt that develops when water precipitates from the frozen solution as ice. By adding a compound to the solution prior to freezing, the concentration of salt attained at any given temperature is reduced. The correlation of salt concentration and cell damage, both in the presence of ice and its absence, was remarkable. However, Meryman countered this argument by proposing that it is not the attainment of a high salt concentration per se that causes the damage but the reduction in cell volume that result from the hypertonic environment.

In a similar fashion, the addition of cryoprotective molecules increases the total volume of the unfrozen fraction of a solution at any given temperature and also reduces the minimum volume attained by a cell during freeze-induced dehydration. This effect has also been proposed as the reason for the colligative protection by these molecules when used during freezing.

It has been suggested that some molecules with cryoprotective properties are able to stabilize macromolecular structure in a reduced liquid water environment by preferentially being excluded from the surface, rendering the surface more hydrated that it otherwise would be in the absence of the molecules. In essence, this reduced the dehydrating effect of the concentrated solution to which the molecules are exposed.

From the early days of cryobiology, ice formation was hypothesized to be a major contributor to cell damage. The chemical effects of ice formation were alluded to above, as the chemical composition of the unfrozen fraction increases due to ice formation. Cells that start off suspended in the solution are sequestered in the unfrozen fraction and are directly exposed to this freeze-concentrated solution. This produces a chemical potential gradient between the intracellular and extracellular solutions, resulting in the exosmosis of cellular water, in an effort to restore chemical potential equilibrium. The result of this is an increase in solute concentration and a decrease in cell volume, which as mentioned have both been proposed to be significant contributors to cell damage.

Cellular damage directly from ice crystals has also been proposed. Mechanical forces from growing ice crystals are thought to play a role in damage when the ice forms inside of cells as well as outside and also outside of cells in organized tissue—the later effect being of particular importance for the current discussion. Other proposals to explain the damaging effects of intracellular ice include intracellular osmotic damage/dilution shock, protein denaturation, and gas bubble formation.

All of the hypothesized mechanisms of cellular damage resulting from attempts at cryopreservation are expected to affect organs as well. The complications arising from trying to optimize cellular preservation are exacerbated by one or more orders of magnitude simply due to the size of the organs that are targets for preservation. Typical isolated cells have a diameter on the order of 5–10 μm, whereas the sizes of organs are measured in centimeters. This presents challenges when it comes to the introduction and removal of cryoprotectants as well as the removal and introduction of heat during the process of preservation.

Heat and Mass Transfer during Organ Cryopreservation

As for isolated cells, it will be necessary to prevent damaging levels of intracellular ice formation in cells within a tissue, as well as within the intercellular spaces. The primary means to accomplish this is by increasing the solute concentration of these compartments by either freeze dehydration or initially loading vitrifiable concentrations of cryoprotectants into the tissue.

While it can take up to several minutes to safely load isolated cells with cryoprotectants during a freezing procedure, the luxury of direct exposure when bathed in a cryoprotectant solution and short diffusion distances from the outside to the inside are not available for cells within an organ. The time required for diffusion to occur scales with the square of the diffusion distance. This challenge is best illustrated by example. The time necessary for a protein molecule to diffuse into the center of a cell is on the order of seconds and to diffuse across a thin piece of tissue (2 mm) is in the order of 27 h. Hence, simply placing an isolated organ into a bath containing cryoprotectants will not suffice to load these molecules into the cells for preservation. As a result, procedures for the cryopreservation of organs usually rely upon perfusing the organ's vasculature with cryoprotectants to significantly reduce the distance necessary for these compounds to travel in order to protect the cells on the interior of the structure.

A similar problem occurs with the need to dehydrate the cells prior to storage at cryogenic temperatures. In a freezing method, one relies upon the water to diffuse out of the cells and precipitate in the extracellular space in order to protect the intact cell from the damaging effects of intracellular ice formation. The same problem of scale exists although for a small molecule like water, the diffusion will take place faster. However, ice formation outside of the cells is not innocuous in a tissue, as has been shown by several investigators. Even with well-intentioned attempts to prevent intracellular ice formation, gradients of cryoprotectant concentrations and osmotic imbalances across a tissue can result in varying levels of stress and post-thaw viability across a tissue, resulting in damage associated with cryoprotectant toxicity and excessive osmotic stress.

Problems associated with heat transfer are analogous to that of mass transfer. Thermal gradients will be established when freezing a bulk solution containing an organ as a result of the heat sink being located on the outside of the vessel, and the necessity for heat to diffuse out of the organ and into this sink. The success of freezing isolated cells as measured by post-thaw viability is known to be strongly dependent upon cooling rate, and this is likely to be similar for cells within the confines of a tissue. Hence, thermal gradients and associated freezing rates may be deleterious to cells located within an organ during a cryopreservation procedure.

Thermal gradients and associated thermo mechanical stress issues are exacerbated in larger systems. This type of stress can actually lead to fracturing of frozen and vitrified solutions, causing significant structural damage to the biomaterial contained within. A relatively simple method to avoid fracturing is to reduce the cooling and warming rates, which reduces the degree to which thermal gradients develop in the system. However, this may result in other changes, such as prolonged exposure to cryoprotectant solutions, and should be used in conjunction with other approaches to optimize a cryopreservation procedure.

It has also been shown that, at least under some conditions, extracellular ice forms preferentially in the vasculature. This is an obvious concern when trying to preserve organs with the goal of transplantation and anastomosis.

Many of the problems associated with freezing organs could, at least in theory, be prevented by the total elimination of ice formation in the sample. Vitrification, as this approach is called, is not a new concept but has its roots in ideas dating back to the 1930s, with Luyet developing an extensive research program in vitrification during this time. Contrary to some of Luyet's work as well as that of modern embryologists, organ vitrification cannot be accomplished using ultra-rapid methods of cooling and warming. It is still possible to vitrify a solution as large as a whole organ, however, given that the solution used for vitrification is designed for this purpose. Furthermore, avoidance of fractures in vitrified samples can be achieved.

This is not to suggest that developing a successful vitrification procedure for whole organs is simple and straightforward. On the contrary, to vitrify at a slow rate of cooling and warming requires solute concentrations on the order of 8 mol/L and a means to introduce the cells to that solution without lethal consequences. Significant research efforts have gone into improving the vitrification solutions and their application used for vitrifying various systems, and the results of these studies suggest that we are approaching the goal of the successful vitrification of some types of organs.

Insight from Nature and Molecular/Genetic Mechanisms

Many mammalian animal systems (such as hibernatory mammals) have become adapted to cold exposure, allowing them to thrive in extremely cold regions where they would otherwise not be able to survive. Nature has evolved several key mechanisms that enable hypothermic and hypoxic survival in animals. Of these mechanisms, the ones that can potentially be most useful for organ preservation include (i) ability to suppress global metabolic rate, (ii) metabolic pathways capable of sustaining and/or delivering minimal essential energy during hypoxia and/hypothermia, and (iii) molecular and cellular defense mechanisms that allow normal metabolism to return upon arousal. Many of these are conserved across animal phyla, which may be why human organs can be cooled to some extent with minimal impact on physiological functionality. However, even if human organs have some of these cold-tolerant genes, they do not necessarily have the capacity to access and apply them, which is why human organs can only tolerate cooling for limited time periods.

Some amphibian and reptile species that live in seasonally cold climates have also developed the ability to survive long-term freezing with about two thirds of their body water frozen as extracellular ice. Surviving this frozen state requires a symphony of responses by cells and organs to sustain homeostasis and allow physiological reconstitution upon warming. This includes directing energy toward the most efficient cellular processes to maintain viability and protection of macromolecules (beyond metabolic production of cryoprotectants, e.g., antioxidant defenses). In this way, freeze tolerance is similar to hypoxia and dehydration, which are also commonly encountered environmental stresses. Freeze/thaw-responsive control of gene expression is now a very active area of research. This epigenetic control of gene expression is likely allowing for anoxia tolerance as well as metabolic suppression and creates unique patterns of freeze-responsive gene and protein expression. This has now been documented via traditional methods, such as signaling pathways impacting transcriptional factors, and also via post-transcriptional regulation of messenger RNA (mRNA) transcripts by small non-coding RNA species that influence the fate of mRNA (microRNAs).

Understanding of the natural mechanisms of animal cold hardiness, hypoxic tolerance, and dehydration is not only key to understanding the principles and mechanisms by which organisms survive cooling to freezing temperatures and rewarming; it may also enable breakthroughs in the banking of human organs and other CVAs. Evidence supports that many of the mechanisms used by these cold-tolerant animals are conserved in hibernating mammals and even primates, with some evidence that analogous mechanisms are found in humans as well.

The few molecular studies of gene expression targeting organ transplant specifically have largely targeted reperfusion injury. These studies have focused on the role of certain signaling pathways such Bcl-2 and Bag-1, two regulatory genes controlling apoptosis to reduce cell damage through hypoxia. There is also a possible effect on the heme oxygenase-1 (HO-1) pathway (a stress protein induced in response to oxidative challenges) and on the hypoxia-inducible factor (HIF) pathway that responds to challenges in available oxygen. Gene therapy has also been proposed as an option to recondition donor lungs ex vivo. Human lungs transfected for IL-10 expression exhibited improved functional quality, increased oxygenation, decreased vascular resistance, improved cell-cell interaction, and a positive shift from a proinflammatory to an antiinflammatory cytokine release.

Proteomic technology in conjunction with enhanced understanding of the epigenetics involved is now allowing identification of proteins related in cellular adaptation in response to I/R injury and has begun to allow investigation into whether these molecular clues might explain the phenotypic changes expressed in the cells.

Many obstacles remain before these studies can be translated into clinical application. With advances in gene therapy, there is promise that such obstacles may be overcome; however, improved methods for vector transfer and delivery protocols (without toxicity) and the timing of gene induction (among many other details) will need to be developed to fully integrate the two fields.

Mesenchymal Stem Cells and Organ Preservation

Adult-derived stem cells such as mesenchymal stem cells (MSCs) offer great therapeutic promise for numerous clinical applications. They can be derived from virtually any tissue in the body that heals, including bone marrow, blood, adipose, dental pulp, umbilical cord/placenta, liver, pancreas, and brain. To date, MSCs originating in bone marrow have arguably been the most extensively studied and are the first cells to be used successfully in therapy. Commonly described as "mesenchymal stem cells" or "mesenchymal stromal cells" (MSCs), these cells have been shown to possess multipotentiality when induced ex vivo. When isolated by adherence to plastic and expanded ex vivo, these cells demonstrate the ability to differentiate into a broad range of cell types of mesodermal origin, including osteoblasts, adipocytes, and chondrocytes, and to some extent ectodermal (neuronal) and endodermal (hepatocytes) origins. In addition to their ability to differentiate into mesodermal tissues, MSCs secrete a variety of cytokines and growth factors that have very potent paracrine activity, distinct from the direct differentiation of MSCs into tissue. These secreted bioactive factors suppress the local immune system, inhibit fibrosis (scar formation) and apoptosis, enhance angiogenesis, and stimulate mitosis and differentiation of tissue-intrinsic reparative or stem cells. Several studies have tested the use of MSCs in models of cardiac damage (infarct), brain damage (stroke), and meniscal regeneration focusing on such paracrine influence (MSC-mediated trophic effects) as the means of tissue repair.

In renal studies, intravenously administrated MSCs have been shown to home to ischemic tubular sites and can be detected within the first hour after cell infusion. The exact process of MSC migration and recruitment is still under debate; however, it is thought to be caused by migration down a gradient of stromal cell-derived factor-1 (SDF-1, a chemokine protein) induced by proinflammatory stimuli followed by adhesion to molecules generated and expressed by injured tissue. Cell-fate and cell-tracking studies have provided support for this theory by demonstrating that MSCs administered exogenously are transiently present at the injury site before being cleared from the circulation. It has also been demonstrated that MSCs are able to actively transmigrate into inflamed tissue across TNF-α-activated endothelium and become partially integrated in the endothelial layer.

MSCs can provide benefit during both ischemia and reperfusion to support the repair of damaged tissue. MSCs release potent growth factors such as vascular endothelial growth factor (VEGF), insulin-like growth factor (IGF), hepatocyte growth factor (HGF), keratinocyte growth factor (KGF), monocyte chemoattractant protein-1 (MCP- 1), SDF-1, and fibroblast growth factor (FGF). These factors stimulate endogenous cellular repair mechanisms, including stimulation of proliferation and angiogenesis. In response to inflammation, it has been demonstrated that MSCs modulate or alter inflammatory responses and release immune-mediators, such as IL-10, IL-6, TGF-b, prostaglandin E2 (PGE2), NO, and IDO and generate a local antiinflammatory prore-parative cellular state. In addition to their antiinflammatory actions, MSCs may induce regulatory T cell expansion which could mitigate allogeneic organ rejection as well as provide support against I/R injury. It has also recently been shown that administration of MSCs promotes macrophages to shift toward the antiinflammatory M2 phenotype, as characterized by expression of a mannose receptor (CD206) and a distinct cytokine profile, consisting of high levels of IL-10 and IL-6 and low levels of IL-12 and TNF-a.

Intravenous infusion of MSCs for acute renal injury has been demonstrated to result in accelerated tubular repair and improved kidney function. In a rat model clamping of the renal pedicles followed by intraaortic infusion of MSCs resulted in enhanced organ function and resulted in fewer apoptotic cells as compared to vehicle-only treated controls. More recently, MSCs were also shown to stimulate tubular cell proliferation, reduce acute tubular necrosis, and suppress oxidative stress and proinflammatory responses or hinder the progression of fibrosis. Other studies have focused on further enhancing the MSC secretome to benefit organ transplantation, including human umbilical cord-derived MSCs designed to overexpress HGF (demonstrated in a rat I/R model) or BMP-7-transduced MSCs (demonstrated in a rabbit model). In rat kidney transplantation studies, MSCs showed promising results and demonstrated less cold ischemia-induced inflammation in early acute allograft rejection as well as long-term benefit in a chronic allograft nephropathy model.

Intravenous infusion of MSCs in endotoxin-induced lung injury models resulted in decreased proinflammatory cytokines, increased antiinflammatory cytokines, and higher survival rates. Additionally, utilizing a gene therapy approach to modify MSCs to deliver IL-10 resulted in reduced signs of injury, lower numbers of apoptotic cells, and less immune cell infiltration in a rat model of lung I/R injury.

Human clinical trials utilizing MSCs to enhance human solid organ transplantation have begun, with several trials registered with the US FDA on clinicaltrials.gov. A phase II trial which has finished (#NCT00658073) compared autologous MSC-based cell therapy to traditional anti-IL-2 receptor antibody induction therapy in patients undergoing renal transplantation and demonstrated lower incidences of acute rejection and decreased risk of opportunistic infection in the MSC-treated group. Another similar phase I/II trial (#NCT00734396), which has also completed, evaluated the safety and feasibility of autologous mesenchymal stem cell therapy in HLA-DR mismatched kidney transplant patients with subclinical rejection (SCR) and or an increase in interstitial fibrosis/tubular atrophy (IF/TA) in the renal biopsy 4 weeks or 6 months after renal transplantation. In that study, the autologous BM MSCs were determined to be clinically feasible and safe, and the findings were suggestive of systemic immunosuppression. Another completed phase II clinical trial utilized human umbilical cord-derived Wharton's jelly as the source of MSCs (#NCT01291329) and investigated intracoronary injection of the MSCs for cardiac regeneration in patients with acute myocardial infarction. Other combined phase I/II clinical trials utilizing intramyocardial administration of autologous bone marrow-derived MSCs to treat severe chronic myocardial ischemia or chronic ischemic ventricular dysfunction secondary to myocardial infarction (#NCT00587990 and #NCT00260338) have also now completed. As more clinical trials are completed and published, a deeper understanding of the therapeutic potential of MSCs will be gained; however, it cannot be denied that the potential for MSCs to ameliorate myocardial ischemia could have profound benefit on organ preservation.

Novel Trophic Factors for I/R Injury

Over the last decade, renewed emphasis has been placed on trophic factors as a means to avoid or mitigate organ damage during transplantation, resulting in a number of very important studies. McNaulty et al. examined the effect of trophic factor supplementation of University of Wisconsin (UW) solution on cold storage of canine kidneys. In that study, bovine neutrophil peptid-1 (bactenecin), substance P, nerve grown factor, epidermal growth factor, and insulin-like growth factor-1 were all added to UW lactobionate solution, and kidneys were stored for up to 4 days. Dogs transplanted with kidneys stored in the supplemented solution had significantly lower peak serum creatinine values and returned to normal faster than kidneys stored for only 3 days in unmodified solutions.

Importantly, while some of the factors used in McNaulty's study, such as IGF-1 and EGF, have the potential to exert beneficial effects by affecting mechanisms of reperfusion injury, this was the first study in which it was conclusively demonstrated that using trophic factors only during ex vivo cold storage could substantially suppress cold ischemic injury. Treating organs in this way resulted in dramatic and immediate effects on function following transplantation. Their results supported the hypothesis that the reduction of available trophic factors due to suppressed metabolism under cold ischemia results in cold ischemic injury during hypothermic preservation using synthetic cold-storage solutions and underscores the potential effect these factors can exert. It was also the first study to demonstrate consistently successful preservation of kidneys for up to 6 days without the use of perfusion technologies. This study showed that trophic factors appear to exert biologic effect in tissues even at low temperatures; however, the mechanism of the influence these factors have on cellular activity during cold storage remains unclear. Previous studies have shown that trophic factors and cell signaling pathways can have significant activity at temperatures as low as 0 °C. However, certain pathways and cascades may have discrete steps that are cold sensitive and shut down at low temperature, and although the overall pathway may be activated in the cold, the actual effector mechanisms may not be realized until the cells are rewarmed. Additionally, expression of mRNA for various trophic factors and peptides, such as vascular endothelial growth factor and endothelin, can be increased during cold storage, while their corresponding receptors may be concurrently downregulated, yielding no net increase in effect. In studies investigating hypothermic storage of lamb cerebral and coronary arteries, the low temperature appeared to increase tyrosine kinase phosphorylation of proteins. These studies support the theory that many cell signaling pathways remain active even at low temperatures and can still be affected by trophic factors. Exactly how these factors affect cell metabolism under hypothermic conditions and how they may buffer reperfusion injury still remain to be determined.

Recent studies have demonstrated that conditioned media from bone marrow MSCs containing trophic factors and microvesicles (MSC-MVs) administered immediately after induction of I/R reverses acute kidney injury (AKI) and protects against chronic kidney disease (CKD). MSC-MVs are reported to contribute to the observed regenerative effects of MSCs in I/R injury by transferring RNA and providing ATP through mitochondrial transfer to ischemic tissue. This restoration of energy supply mediated by MSC-MVs is crucial to drive repair of ischemia-damaged tissues and may account for the beneficial effects of administration of MSCs in early stage I/R injury. Prosurvival gene activation has also been demonstrated through MSC-MVs. In a study by Bruno et al., MSC-MVs were shown to induce the prosurvival genes (Bcl-xL, Bcl2, and BIRC8) and downregulate proapoptotic genes like Casp1, Casp8, and LTA.

The effects of MSC conditioned media (MSC-CM) trophic factors were shown to be associated with a transient recruitment of MSCs within the kidney with minimal cellular incorporation into the regenerating tubules. Based on this observation, it has been suggested that MSCs may provide a paracrine support to kidney repair. This concept is supported by other studies involving intravenous treatment with MSC-CM. In one study, capillary density and cardiac function were both improved following IV administration of human ESC-derived MSC-CM following myocardial infarction (MI) in a mouse model of myocardial I/R injury. In another study utilizing a pig model of MI, human MSC-CM administered IV was shown to increase capillary density and preserve cardiac function, likely by increasing myocardial perfusion.

Consistent with the above reports, Bi et al. demonstrated that administration of MSC-CM may mimic the beneficial effects of the stem cell therapy, indicating that engraftment of MSCs in the renal tubules is not necessary, and Bruno et al. reported that intravenous administration of MVs derived from human MSCs has the same efficacy of MSCs on the functional and morphological recovery of glycerol-induced AKI in severe combined immunodeficient mice.

Other novel sources of trophic factors to assist in organ repair/recovery following I/R injury are also now being investigated. Ongoing work by Petrenko focusing on fetal liver stem cells (both hepatic and hematopoietic) has yielded trophic factor "cocktails" which can be utilized as cell-free extracts. Petrenko has previously shown significant therapeutic effects of these trophic factors (TFs) on models of acute toxic hepatitis, experimental cirrhosis, chronic alcohol poisoning, solid tumor progression, and wound healing. Most recently, utilizing a non-hepatic source of mesenchymal stromal cell-derived cell-free fetal-specific factors, Petrenko was able to demonstrate that supplementation of preservation solutions with such factors modulated redox-dependent processes and led to strengthening of cell adaptive responses to stress against I/R injury. These results indicate the potential for TF derived from adult stem/progenitor cells and or conditioned culture media may be useful in reducing organ I/R injury.

The protective effect of MSC trophic factors seems to be supported by numerous studies; however, their specific mechanism(s) of action still remain unclear. These protective effects have been referred to as "survival factors" in cell culture studies and are often associated with antiapoptotic effects, protection against mitochondrial injury, and overall enhanced cell growth. One potential mechanism is modulation of ERK1/2, p38 MAPK, and HO-1 signaling pathways during hypothermia and rewarming. While preliminary, these studies have demonstrated that modulating these pathways is correlated with reduced graft injury and improved long-term viability. These trophic factors also seem to affect chemokine production and therefore leukocyte/pericyte recruitment, and modulation of these signaling pathways through hypothermic storage has been described; however, means to exploit this favorably under clinical storage conditions are yet to be described.

References

- Cryonics, life-genetic: science.howstuffworks.com, Retrieved 18 March, 2019
- Daley ME, Spyracopoulos L, Jia Z, Davies PL, Sykes BD (April 2002). "Structure and dynamics of a beta-helical antifreeze protein". Biochemistry. 41 (17): 5515–25. doi:10.1021/bi0121252. PMID 11969412
- Brown DJ, Brugger H, Boyd J, Paal P (November 2012). "Accidental hypothermia". The New England Journal of Medicine. 367 (20): 1930–8. doi:10.1056/NEJMra1114208. PMID 23150960
- Van Den Broeck, U.; Vandermeeren, M.; Vanderschueren, D.; Enzlin, P.; Demyttenaere, K.; d'Hooghe, T. (2012). "A systematic review of sperm donors: Demographic characteristics, attitudes, motives and experiences of the process of sperm donation". Human Reproduction Update. 19 (1): 37–51. doi:10.1093/humupd/dms039. PMID 23146866
- Biology of cell survival in the cold: the basis for biopreservation of tissues and organs. Advances in Biopreservation. J. G. a. B. Baust, J.M. Boca Raton, Taylor and Francis Group: 15–62, 2007

PERMISSIONS

All chapters in this book are published with permission under the Creative Commons Attribution Share Alike License or equivalent. Every chapter published in this book has been scrutinized by our experts. Their significance has been extensively debated. The topics covered herein carry significant information for a comprehensive understanding. They may even be implemented as practical applications or may be referred to as a beginning point for further studies.

We would like to thank the editorial team for lending their expertise to make the book truly unique. They have played a crucial role in the development of this book. Without their invaluable contributions this book wouldn't have been possible. They have made vital efforts to compile up to date information on the varied aspects of this subject to make this book a valuable addition to the collection of many professionals and students.

This book was conceptualized with the vision of imparting up-to-date and integrated information in this field. To ensure the same, a matchless editorial board was set up. Every individual on the board went through rigorous rounds of assessment to prove their worth. After which they invested a large part of their time researching and compiling the most relevant data for our readers.

The editorial board has been involved in producing this book since its inception. They have spent rigorous hours researching and exploring the diverse topics which have resulted in the successful publishing of this book. They have passed on their knowledge of decades through this book. To expedite this challenging task, the publisher supported the team at every step. A small team of assistant editors was also appointed to further simplify the editing procedure and attain best results for the readers.

Apart from the editorial board, the designing team has also invested a significant amount of their time in understanding the subject and creating the most relevant covers. They scrutinized every image to scout for the most suitable representation of the subject and create an appropriate cover for the book.

The publishing team has been an ardent support to the editorial, designing and production team. Their endless efforts to recruit the best for this project, has resulted in the accomplishment of this book. They are a veteran in the field of academics and their pool of knowledge is as vast as their experience in printing. Their expertise and guidance has proved useful at every step. Their uncompromising quality standards have made this book an exceptional effort. Their encouragement from time to time has been an inspiration for everyone.

The publisher and the editorial board hope that this book will prove to be a valuable piece of knowledge for students, practitioners and scholars across the globe.

INDEX

A

Aerosol Sprays, 45

Androgen Deprivation Therapy, 38

Arachidonic Acid, 17

Artificial Insemination, 2, 87-88, 91, 187, 193

Atrial Fibrillation, 180, 183

B

Body Mass Index, 183

C

Cell Dehydration, 13, 66, 92

Cell Monolayers, 16

Cell Shrinkage, 14-15, 96

Cellular Collapse, 2

Cervical Cancer, 28

Chemotherapy, 28-30, 35-36, 39-41, 48-49, 101, 103, 110, 119, 193

Cold Spray, 45

Cooling Rate, 4, 6, 11, 14-16, 53-54, 57-58, 60, 62, 64-67, 71-72, 76, 88, 94, 110-113, 115, 205

Cryoablation, 12, 30-44, 47

Cryoablation Cycle, 36

Cryoimmunotherapy, 40-41

Cryomapping, 33

Cryoneurolysis, 12, 42-43, 172

Cryopreservation, 1-2, 10-11, 14, 48-50, 52-53, 59-63, 65-66, 68, 75, 77, 79, 82-83, 86-97, 100-111, 113, 115-116, 118-125, 127-128, 177, 203-205

Cryoprobes, 23, 30

Cryosurgery, 12, 14-15, 18-21, 23-30, 33, 38, 44-47, 177

Cytotoxic T Cells, 18, 34, 39

D

Dendritic Cells, 34-35, 38

Dermatofibroma, 19, 24, 26

Diethyl Ether, 8

Dimethyl Sulfoxide, 7-10, 48, 93

Direct Cell Injury, 15, 17

E

Electron Transport Chain, 17

Endothelial Cells, 18

Endothelial Damage, 17

Ethylene Glycol, 7, 9-10, 53, 62, 93, 101, 109, 147-148, 171, 173

Eukaryotic Cells, 7, 11

External Beam Radiation Therapy, 31

F

Freezing Point Depression, 6

Frostbite, 16-18, 46, 179

G

Glass Transition Temperature, 3, 58, 62-64, 71, 112, 116, 145

H

Heat Insulation, 182

Heat Shock Protein, 34

Hormone Therapy, 28, 30

Hydroxyl Group, 8

Hypothermia, 13, 49-50, 61, 177-187, 201, 206, 211-212

I

Ice Crystallization, 11, 93

Immunological Injury, 14, 18

Immunotherapy, 33-41

Ionic Composition, 13, 200

L

Leukocytes, 18, 202

Lipid Peroxidation, 17, 202

Liquid Nitrogen, 1, 19-25, 27, 43, 45-46, 48, 55-58, 64-66, 70-72, 75, 78, 83, 86-87, 93, 95, 97, 101, 108, 117, 124, 131, 134, 138, 148, 150, 192, 196

M

Mammalian Cells, 13, 50

Melanocytes, 15, 21, 27

Melanoma, 18, 26, 40

Metabolic Processes, 2, 59, 116

Myelin Sheath, 42

N
Natural Killer, 18, 34-36
Neutrophil Activation, 17-18
Nitrous Oxide, 42-44

O
Osmosis, 6, 11, 15, 83
Osmotic Pressure, 6, 16, 52, 67, 173
Osmotic Stress, 11, 52, 57, 205

P
Prostate Cancer, 15, 28-29, 37-38, 41, 47
Pulmonary Edema, 180, 183

R
Radiation Therapy, 28-29, 31, 37, 39, 101
Radical Retropubic
Prostatectomy, 31
Renal Cell Carcinoma, 31, 37

S
Seborrheic Keratosis, 19, 25
Spot Freeze Technique, 19, 21-23
Spray Methods, 19, 21
Stereoisomers, 10
Sun-damaged Skin, 24

T
T Cell Receptor, 36-37
Target Lesion, 22, 24
Temperature Gradient, 14, 64, 117
Thawing Process, 15, 102, 109, 122, 190
Thermal Energy, 44, 115
Thermal History, 13-14
Thermal Insulation, 161, 182
Timed Spot Freeze Technique, 19, 21-23
Tumour Cells, 14-15, 18, 34-37, 49, 99

V
Vapor Pressure, 6
Vascular Stasis, 17
Vitrification, 3-5, 7, 10-11, 50, 56-59, 62, 66, 68-69, 71-72, 79, 81, 83-84, 93-96, 101-102, 104-106, 109, 111-113, 115, 125-126, 128, 205-206

X
Xanthine Oxidase, 17-18, 202

www.ingramcontent.com/pod-product-compliance
Lightning Source LLC
LaVergne TN
LVHW080605210926
848485LV00003B/106

* 9 7 8 1 6 4 7 4 0 0 9 8 9 *